17
Topics in Heterocyclic Chemistry

Series Editor: R. R. Gupta

Editorial Board:
D. Enders · S. V. Ley · G. Mehta · K. C. Nicolaou
R. Noyori · L. E. Overman · A. Padwa

Topics in Heterocyclic Chemistry
Series Editor: R. R. Gupta

Recently Published and Forthcoming Volumes

Heterocyclic Supramolecules I
Volume Editor: K. Matsumoto
Volume 17, 2008

Bioactive Heterocycles VI
Flavonoids and Anthocyanins in Plants,
and Latest Bioactive Heterocycles I
Volume Editor: N. Motohashi
Volume 15, 2008

Heterocyclic Polymethine Dyes
Synthesis, Properties and Applications
Volume Editor: L. Strekowski
Volume 14, 2008

Synthesis of Heterocycles via Cycloadditions II
Volume Editor: A. Hassner
Volume 13, 2008

Synthesis of Heterocycles via Cycloadditions I
Volume Editor: A. Hassner
Volume 12, 2008

Bioactive Heterocycles V
Volume Editor: M. T. H. Khan
Volume 11, 2007

Bioactive Heterocycles IV
Volume Editor: M. T. H. Khan
Volume 10, 2007

Bioactive Heterocycles III
Volume Editor: M. T. H. Khan
Volume 9, 2007

Bioactive Heterocycles II
Volume Editor: S. Eguchi
Volume 8, 2007

Heterocycles from Carbohydrate Precursors
Volume Editor: E. S. H. El Ashry
Volume 7, 2007

Bioactive Heterocycles I
Volume Editor: S. Eguchi
Volume 6, 2006

Marine Natural Products
Volume Editor: H. Kiyota
Volume 5, 2006

**QSAR and Molecular Modeling Studies
in Heterocyclic Drugs II**
Volume Editor: S. P. Gupta
Volume 4, 2006

**QSAR and Molecular Modeling Studies
in Heterocyclic Drugs I**
Volume Editor: S. P. Gupta
Volume 3, 2006

Heterocyclic Antitumor Antibiotics
Volume Editor: M. Lee
Volume 2, 2006

Microwave-Assisted Synthesis of Heterocycles
Volume Editors: E. Van der Eycken, C. O. Kappe
Volume 1, 2006

Heterocyclic Supramolecules I

Volume Editor: Kiyoshi Matsumoto

With contributions by
K. Goto · S. Inokuma · K. Ishibashi · M. Ito · N. Komatsu · S. Mameri
J. Nishimura · T. Okujima · N. Ono · S. Shinoda · R. Tamura
H. Tsue · H. Tsukube · H. Yamada

The series *Topics in Heterocyclic Chemistry* presents critical reviews on "Heterocyclic Compounds" within topic-related volumes dealing with all aspects such as synthesis, reaction mechanisms, structure complexity, properties, reactivity, stability, fundamental and theoretical studies, biology, biomedical studies, pharmacological aspects, applications in material sciences, etc. Metabolism will be also included which will provide information useful in designing pharmacologically active agents. Pathways involving destruction of heterocyclic rings will also be dealt with so that synthesis of specifically functionalized non-heterocyclic molecules can be designed.

The overall scope is to cover topics dealing with most of the areas of current trends in heterocyclic chemistry which will suit to a larger heterocyclic community.

As a rule contributions are specially commissioned. The editors and publishers will, however, always be pleased to receive suggestions and supplementary information. Papers are accepted for *Topics in Heterocyclic Chemistry* in English.

In references *Topics in Heterocyclic Chemistry* is abbreviated *Top Heterocycl Chem* and is cited as a journal.

Springer WWW home page: springer.com
Visit the THC content at springerlink.com

ISBN 978-3-540-68189-2   e-ISBN 978-3-540-68192-2
DOI 10.1007/978-3-540-68192-2

Topics in Heterocyclic Chemistry ISSN 1861-9282

Library of Congress Control Number: 2008926598

© 2008 Springer-Verlag Berlin Heidelberg

This work is subject to copyright. All rights are reserved, whether the whole or part of the material is concerned, specifically the rights of translation, reprinting, reuse of illustrations, recitation, broadcasting, reproduction on microfilm or in any other way, and storage in data banks. Duplication of this publication or parts thereof is permitted only under the provisions of the German Copyright Law of September 9, 1965, in its current version, and permission for use must always be obtained from Springer. Violations are liable to prosecution under the German Copyright Law.

The use of general descriptive names, registered names, trademarks, etc. in this publication does not imply, even in the absence of a specific statement, that such names are exempt from the relevant protective laws and regulations and therefore free for general use.

Cover design: WMXDesign GmbH, Heidelberg
Typesetting and Production: le-tex publishing services oHG, Leipzig

Printed on acid-free paper

9 8 7 6 5 4 3 2 1 0

springer.com

Series Editor

Prof. R. R. Gupta†

10A, Vasundhara Colony
Lane No. 1, Tonk Road
Jaipur-302 018, India
rrg_vg@yahoo.co.in

Volume Editor

Prof. Dr. Kiyoshi Matsumoto

Elm Institute of Science
Hiei-dira 2-32-36
Otsu 520-0016, Japan
Kiyoshi.Matsumoto@ma1.seikyou.ne.jp

Editorial Board

Prof. D. Enders

RWTH Aachen
Institut für Organische Chemie
D-52074, Aachen, Germany
enders@rwth-aachen.de

Prof. Steven V. Ley FRS

BP 1702 Professor
and Head of Organic Chemistry
University of Cambridge
Department of Chemistry
Lensfield Road
Cambridge, CB2 1EW, UK
svl1000@cam.ac.uk

Prof. G. Mehta FRS

Director
Department of Organic Chemistry
Indian Institute of Science
Bangalore- 560 012, India
gm@orgchem.iisc.ernet.in

Prof. K.C. Nicolaou

Chairman
Department of Chemistry
The Scripps Research Institute
10550 N. Torrey Pines Rd.
La Jolla, California 92037, USA
kcn@scripps.edu
and
Professor of Chemistry
Department of Chemistry and Biochemistry
University of California
San Diego, 9500 Gilman Drive
La Jolla, California 92093, USA

Prof. Ryoji Noyori NL

President
RIKEN (The Institute of Physical and Chemical Research)
2-1 Hirosawa, Wako
Saitama 351-0198, Japan
and
University Professor
Department of Chemistry
Nagoya University
Chikusa, Nagoya 464-8602, Japan
noyori@chem3.chem.nagoya-u.ac.jp

Prof. Larry E. Overman

Distinguished Professor
Department of Chemistry
516 Rowland Hall
University of California, Irvine
Irvine, CA 92697-2025
leoverma@uci.edu

Prof. Albert Padwa

William P. Timmie Professor of Chemistry
Department of Chemistry
Emory University
Atlanta, GA 30322, USA
chemap@emory.edu

Topics in Heterocyclic Chemistry
Also Available Electronically

For all customers who have a standing order to Topics in Heterocyclic Chemistry, we offer the electronic version via SpringerLink free of charge. Please contact your librarian who can receive a password or free access to the full articles by registering at:

springerlink.com

If you do not have a subscription, you can still view the tables of contents of the volumes and the abstract of each article by going to the SpringerLink Homepage, clicking on "Browse by Online Libraries", then "Chemical Sciences", and finally choose Topics in Heterocyclic Chemistry.

You will find information about the

– Editorial Board
– Aims and Scope
– Instructions for Authors
– Sample Contribution

at springer.com using the search function.

Color figures are published in full color within the electronic version on SpringerLink.

Preface

A wide range of fields within supramolecular chemistry are of current and great interest ranging from nanosciences, medicinal sciences, biosciences, and even organic sciences and this is a mature and extremely active area of research. In 1978, Lehn defined this chemistry as the "chemistry of molecular assemblies and of the intermolecular bond." In other words, supramolecular chemistry is noncovalent chemistry based upon covalent chemistry.

On the other hand, it is well known that replacing the carbon atom of cyclic compounds can lead to dramatic changes in chemical and physical properties and the principles of homocyclic chemistry are often of limited value and may even lead to incorrect results. This is often indeed the case in supramolecular chemistry. The modern explosion of nonochemistry is highly based upon the fundamental recognition of intermolecular interactions engendered by supramolecular scientists.

In this volume entitled *Heterocyclic Supramolecules* I, a part of the series *Topics in Heterocyclic Chemistry*, some selected topics in noncovalent chemistry from the last decade are highlighted, with attention particularly focused on heterocyclic supramolecules as well as heterocycle-based nanosciences.

The first chapter, "Molecular Recognition with designed Heterocycles and their Lanthanide Complexes" by S. Mameri, S. Shinoda, and H. Tsukube describes various synthetic receptors for specific binding of cationic anionic guests mainly in the solution states. Furthermore, special attention is directed at the heterocycle-lanthanide complexes that worked as luminescent sensory devices of biologically important anions. Thus, "rare" earth metals are making the change into "hopeful" earth metals.

The second chapter, "Syntheses and Properties of Crownophanes" by S. Inokuma, M. Ito, and J. Nishimura reviews a variety of crownophanes possessing both crown ether and cyclophane moieties, the latter ranging from benzene to condensed aromatic and heteroaromatic rings whose selective complexation in the solution states principally toward the metal cations are reviewed. The related rotaxanes and catenanes are also described in this chapter.

The third chapter, "Azacalixare: A New Class of Calixarene Family" by H. Tsue, K. Ishibashi, and Rui Tamura presents recent developments in syntheses, conformations, and inclusion properties of nitrogen-bridged calixarene derivatives possessing a $[1_n]$metacyclophane unit. Since just the replacement of hydrogen(s) of the methylene bridge with an appropriate group(s) would offer wider functional variations as in the case of the crownophane family, further developments in this field are surely anticipated.

The fourth chapter, "Chemistry of Calixfurans" by Kei Goto presents summaries of the synthesis, reactions, structures, and host–guest chemistry of calix[n]furans and their hybrid systems containing other aromatic units like pyrrole and thiophene. Calixfurans appear to be a tactful supporting actor in the chemistry of calixarenes. Regardless of their rather modest intrinsic binding abilities, the weak coordination by the furan units of calixfurans or hybrid systems plays a crucial role in certain cases. More importantly, calixfurans can be converted into a wide variety of macrocycles including those otherwise difficult to access since the furan unit serves as a versatile functional group such as a masked 1,4-dicarbonyl equivalent and Diels–Alder diene. Further development of the synthetic strategy of calixfurans as well as the novel methods for their transformation to other functional molecules is highly anticipated. Since the conformational behavior of calixfurans has not been sufficiently clarified yet, the more sophisticated strategy for regulation of their conformational dynamics should be explored for the ready construction of the desired molecular framework.

The fifth chapter, "Supramolecules based on Porphyrins" by H. Yamada, T. Okujima, and N. Ono presents a review particularly focusing on the supramolecular architectures of porphyrins that enable their use as electronic and optical functional materials such as third-order optical susceptibilities, photoenergy conversion systems, and organic field-effect transistors. Although life as we know it would be impossible without porphyrins, they are now characterized not only as the "color of life" but also as a treasure house of material sciences. For instance, photovoltaic cells are currently of broad interest as potential low-cost approaches to solar energy conversion. Large-area electronic devices and solution-processed organic semiconductors based on porphyrins, phthalocyanines, and related molecules could have potentially a huge cost advantage over Si-based devices if conversion efficiency and durability can be improved to the level of Si-solar cells.

The final chapter, "Heterocyclic Supramolecular Chemistry of Fullerenes and Carbon Nanotubes" by N. Komatsu presents an extremely unique review that focuses on the noncovalent chemistry of fullerenes and carbon nanotubes with nitrogen- and/or oxygen-containing heterocyclic molecules such as porphyrin, DNA, protein, peptide, and carbohydrate. Not only exohedral but also endohedral functionalization is reviewed, because the above guest molecules can interact with both faces of the carbon nanotubes. The hurdles in structural separation, nanofabrication, and bioapplications of carbon nanotubes will hopefully be addressed by the supramolecular strategy.

Finally, I hope, in the near future, that heterocyclic supramolecules could figure in a practical generation of molecular machines and in highly effective production of useful materials at molecular levels, for example in a much more efficient artificial photosynthetic process than the natural one and in an electronics revolution that will produce the carbon-heteroatom-based molecular computer—probably more than 1000-times smaller and a million-times more powerful than our present machines, via recourse to quantum mechanics rather than classical Newton mechanics that would solve environmental as well as energy problems.

Kyoto, May 2008   Kiyoshi Matsumoto

Contents

Molecular Recognition with Designed Heterocycles
and Their Lanthanide Complexes
S. Mameri · S. Shinoda · H. Tsukube   1

Syntheses and Properties of Crownophanes
S. Inokuma · M. Ito · J. Nishimura 43

Azacalixarene: A New Class in the Calixarene Family
H. Tsue · K. Ishibashi · R. Tamura 73

Chemistry of Calixfurans
K. Goto 97

Supramolecules Based on Porphyrins
H. Yamada · T. Okujima · N. Ono 123

Heterocyclic Supramolecular Chemistry
of Fullerenes and Carbon Nanotubes
N. Komatsu 161

Subject Index 199

Molecular Recognition with Designed Heterocycles and Their Lanthanide Complexes

Samir Mameri · Satoshi Shinoda · Hiroshi Tsukube (✉)

Department of Chemistry, Graduate School of Science, Osaka City University, 3-3-138 Sugimoto, Sumiyoshi-ku, 558-8585 Osaka, Japan
tsukube@sci.osaka-cu.ac.jp

1	Introduction	2
2	**Cation Recognition with Designed Heterocycles**	6
2.1	Heterocycles for Cation Recognition	6
2.2	Geometry Optimized Receptors	7
2.3	Chirality Optimized Receptors	9
2.4	Designed Heterocycles for Supramolecular Cation Recognition	11
3	**Anion Recognition with Designed Heterocycles**	14
3.1	Heterocycles for Anion Recognition	14
3.2	Cationic Heterocyclic Receptors	17
3.3	Neutral Heterocyclic Receptors	20
3.4	Anion Detection with Designed Heterocycles	25
4	**Anion Recognition with Designed Heterocycle–Lanthanide Complexes** . .	27
4.1	Heterocycle–Lanthanide Complexes for Anion Recognition and Sensing . .	28
4.2	Luminescent Lanthanide Complexes	28
4.3	Lanthanide Complexes for Luminescence Sensing	29
4.3.1	Anion Sensing in Organic Media	29
4.3.2	Anion Sensing in Aqueous Media	32
4.4	Lanthanide Complexes for NMR Sensing	35
5	Conclusion	38
	References	38

Abstract Molecular recognition with designed heterocycles and their lanthanide complexes in the solution states was mainly described. Various synthetic receptors for specific binding of cationic and anionic guests were presented, in which several weak interactions were combined to fit the size, shape, geometry, and electronic characteristics of the specific guest species. The cation-ligating heterocycles were successfully organized in the receptor molecules to exhibit high selectivity and efficiency in the cation recognition and sensing. A series of N-protonated and substituted heterocycles had the potentials as anion receptors effective in aqueous media. Furthermore, the designed heterocycle-lanthanide complexes worked as luminescent sensory devices of biologically important anions. The examples presented here clearly indicated that a variety of heterocycles acted as useful building blocks in the receptor architecture. The sophisticated molecular synthesis using potential heterocycles can provide specific recognition at the molecular and supramolecular levels.

Keywords Chirality · Lanthanide complex · Luminescence · Molecular recognition · Receptor

Abbreviations

AMP	Adenosine 5′-monophosphate
ADP	Adenosine 5′-diphosphate
ATP	Adenosine 5′-triphosphate
CD	Circular dichroism
CPL	Circularly polarized luminescence
DFT	Density functional theory
ESI-MS	Electrospray ionization mass spectrometry
IR	Infrared
MRI	Magnetic resonance imaging
NMR	Nuclear magnetic resonance

1
Introduction

Heterocycles are widely employed as useful building blocks in the construction of molecular recognition and supramolecular assembly systems [1]. These provide hydrogen bondings, electrostatics, metal coordination bonds, π–π interactions, and other attractive weak forces with various species both in the solution and the solid states. Since several families of heterocycles further exhibit electrochemical activities, photochemical reactivities, optical characteristics, and other functions, current heterocyclic chemistry offers a robust basis for sophisticated molecular architectures toward molecular recognition and supramolecular assembly.

There are two types of molecular architecture approaches based on heterocyclic chemistry in this research field: the "convergent" approach that mainly targets molecular recognition; and the "divergent" one that leads to supramolecular assembly. As illustrated in Fig. 1, oligopyridine derivatives were often used in these two approaches. Sauvage et al. successfully employed 1,10-phenanthroline derivatives in the metal-templated synthesis of catenane **1**. In this case, two 2,2′-bipyridine subunits cooperatively coordinated to the same copper(I) center, and a subsequent ring-closure reaction smoothly occurred in a pseudo-intramolecular fashion [2]. Fujita et al. used 4,4′-bipyridine derivatives in the synthesis of metallocycle **2**, so that each pyridine moiety worked independently [3]. When *cis*-protected Pd^{2+} species were complexed with the pyridine derivatives, a large-membered metallocycle **2** was derived from four 4,4′-bipyridines and four Pd^{2+} centers. Both examples indicate promising possibilities that metal-coordinative heterocycles are applicable in the development of molecular recognition and supramolecular assembly systems.

Complementary hydrogen bondings are also involved in many natural and artificial recognition and assembly systems. The photosynthetic reaction cen-

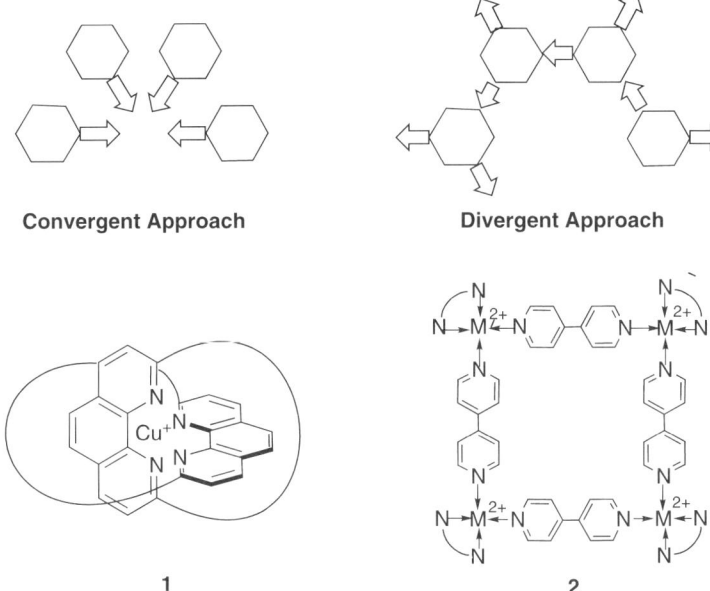

Fig. 1 Convergent and divergent approaches based on heterocyclic chemistry

ter, tetrameric hemoglobin proteins, polyketide synthases, and viral coat assemblies are typical biological examples. Figure 2 illustrates other interesting examples for biological anion recognition, in which complementary hydrogen bondings play crucial roles. In the Cl⁻ anion channel of *Salmonella typhimurium* (see **3**), the Cl⁻ anion is held in place by two O–H···Cl⁻ hydrogen bonds and two N–H···Cl⁻ hydrogen bonds [4, 5]. Several synthetic anion recognition systems have similarly been constructed based on complementary hydrogen bondings with heterocyclic compounds. Urea-functionalized receptor **4** was built up on the cholapod scaffold to exhibit high Cl⁻ anion affinity. In this system, the multiple hydrogen bondings were provided

Fig. 2 Schematic illustration of Cl⁻ anion channel of *Salmonella typhimurium*

4

5

6

7

from three urea units to enhance the Cl⁻ complex stability [6]. Davis et al. demonstrated that synthetic guanosine derivatives formed hydrogen-bonded quartet structures in the solution states. Since the resulting quartets **5** were further stabilized by complexation with alkali metal cations, they acted as self-assembled receptors of the metal cations [7].

In addition to metal coordination bonds, hydrogen bondings, electrostatics, and $\pi-\pi$ interactions, the attractive interaction between the curved π-surface of a fullerene and the flat π-surface of a porphyrin was recently reported [8, 9]. When the fullerene and the porphyrin were linked covalently in the single molecule (see **6**), the two functions interacted with each other to form a supramolecular assembly in the solid state. In contrast to this divergent system, macrocycle **7** was prepared from the fused zinc porphyrins for the convergent interaction with specific guests. Since **7** had space to accommodate two fullerenes, this kind of interaction operated well in the supramolecular assembly and molecular recognition systems.

Other sophisticated examples of molecular recognition were reported with helical foldamers [10–12]. Foldamers **8** spontaneously formed helical structures through intramolecular solvophobic interactions, in which some guests were nicely accommodated. Since the pyridine-containing foldamers **8b** and

8a   R = -CO$_2$(CH$_2$CH$_2$O)$_3$CH$_3$

8b

8c   R = -(C$_2$H$_4$O)$_8$CH$_3$

8c provided effective hydrogen bondings with water and sugar derivatives, the combination of potential heterocycles with three-dimensional helical structures offered unique molecular recognition phenomena.

This chapter focuses primarily on the molecular recognition with designed heterocycles and their lanthanide complexes in the solution state. We present various receptors specific for cationic or anionic guests, in which several weak interactions were optimized to fit the size, shape, geometry, and electronic characters of target guest species. The sophisticated molecular architecture using heterocyclic building blocks was demonstrated to provide specific recognition at the molecular and supramolecular levels. Although many interesting examples have been reported to cover the divergent approaches towards supramolecular assembly systems, we limit here our efforts to highlight molecular recognition with heterocycles and their lanthanide complexes. Several chapters of this volume deal with other interesting aspects of heterocyclic supramolecules, and readers are recommended to refer to them.

2
Cation Recognition with Designed Heterocycles

Several effective strategies have been established based on heterocyclic chemistry to develop the cation-selective receptors. The incorporation of cation-binding heterocycles into the receptor skeleton modifies cation complexation behavior and tunes cation selectivity for specific guests. Furthermore, we can learn many things from biological cation recognition phenomena. For example, lasalocid, valinomycin, and other naturally occurring ionophores well recognize the spherical alkali metal cations. Although these have too large cyclic skeletons and/or too many asymmetric carbons to target the simple metal cations, the following two strategies work well to offer high selectivity and efficiency in biological cation recognition: (1) optimization of the ligand geometry for the target cations, and (2) introduction of the chirality into the receptor. Because each guest cation has shape and coordination characteristics, both strategies are also effective in the development of artificial receptors for cation recognition. After presenting some examples of the heterocycle-based receptors, these two strategies are detailed in this chapter. Further interesting extensions of cation recognition phenomena at the supramolecular level are also addressed.

2.1
Heterocycles for Cation Recognition

Since oxygen, nitrogen, sulfur, and other heteroatoms exhibit characteristic cation coordination natures, incorporation of potential heterocycles into the receptor molecule provides a simple but effective synthetic strategy for

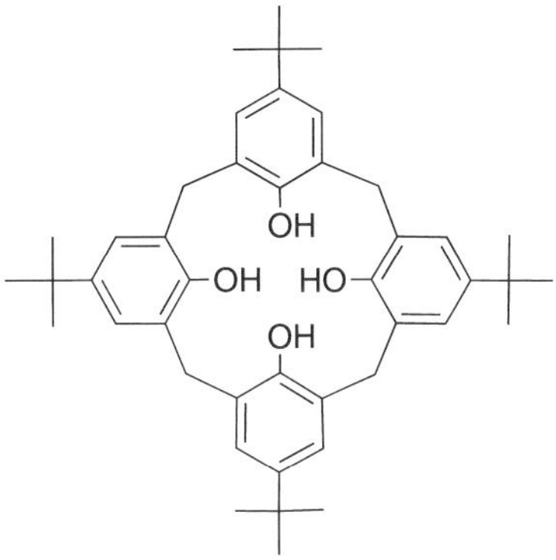

9

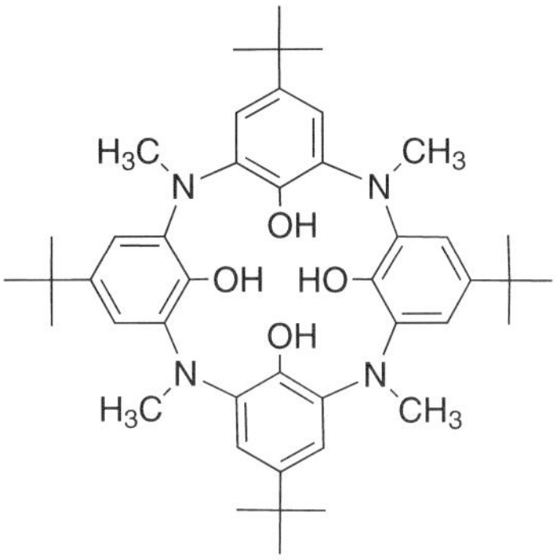

10

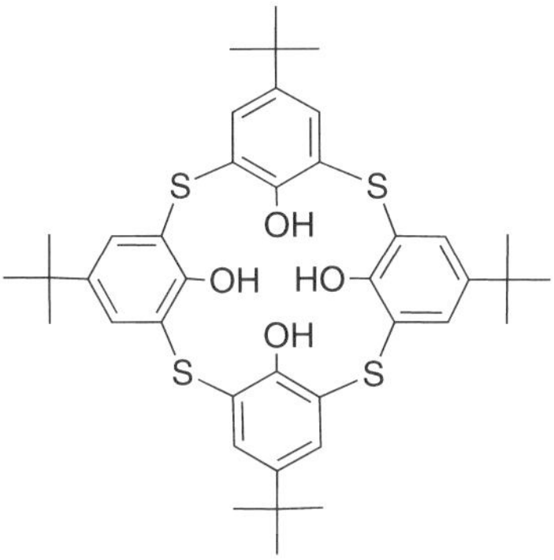

11

the development of cation-specific receptors. In addition to various heterocrown ethers and related macrocycles [1], several heterocalixarenes were presented [13–15]. The original type of calix[4]arene **9** is a *meta*-cyclophane elaborating a cyclic array of phenolic rings joined at the 1,3-positions by methylene bridges. Heterocalix[4]arenes **10** and **11** were typically designed to replace the bridging methylene groups by heteroatoms. Since they exhibited tunable cation-selectivity profiles depending on the donor combinations and the ring sizes, the designed heterocalixarenes displayed new interesting cation recognition functions.

2.2
Geometry Optimized Receptors

As frequently observed with naturally occurring ionophores, geometry optimization of the cation-ligating donor array offered specific cation recognition in artificial receptor systems [16]. The number of successful examples of acyclic systems is still limited, but a series of Ag^+ cation-selective podands **12**

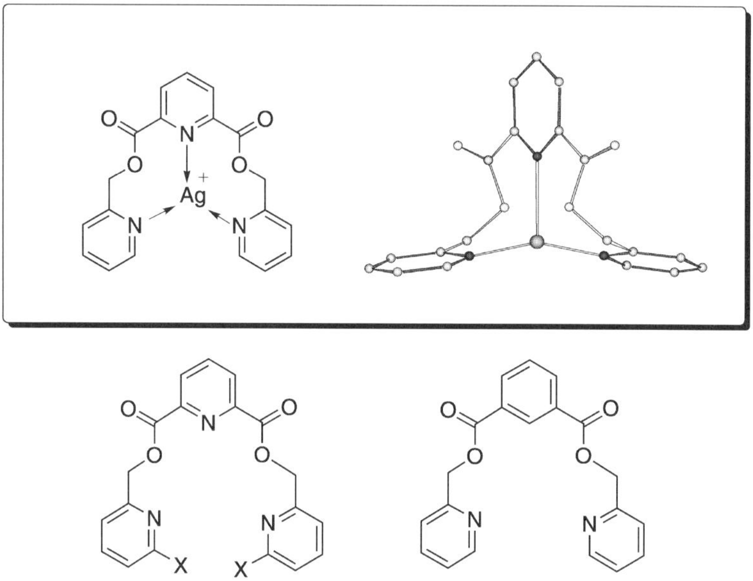

12 **13**

a: X = H
b: X = $CH_2OTBDMS$
c: X = CH_2OPal*
d: X = Br
e: X = CH_2OBn
f: X = CH_2OTr

*Pal = $CO(CH_2)_{14}CH_3$

Fig. 3 Tridentate receptors **12** for selective Ag^+ cation recognition. Reprinted with permission from [17]. © (1998), American Chemical Society

have been developed along this line (Fig. 3) [17]. The Ag^+ cation usually has a linear bidentate coordination mode, but occasionally forms tricoordinated complexes. To develop a tridentate ligand specific for Ag^+ cation, a series of podands **12** were designed to accommodate the Ag^+ cation via cooperative coordination from three pyridine nitrogen atoms. As illustrated in Fig. 3, our calculation using Spartan SGI version 4.0.1 (*ab initio*, STO-3G) suggested tridentate complexation between Ag^+ cation and podand **12a.** ^{13}C NMR titration experiments revealed that the three pyridine rings cooperatively bound the Ag^+ cation. The ester oxygen atoms were located close to the Ag^+ cation, but their lone pair electrons did not point to it. Podand **13** having two pyridine moieties formed a 1 : 1 Ag^+ cation complex, in which the bidentate coordination occurred. Such Ag^+ cation-selective receptors are of great utility in ^{111}Ag-based radioimmunotherapy and photographic techniques as well as in the separation of Ag^+ cation from a natural source or wastewater. The competitive liquid-liquid extraction experiments of Ag^+ cation with Pb^{2+}, Cu^{2+},

Ni^{2+}, Co^{2+}, and Zn^{2+} cations revealed that podands **12** efficiently extracted Ag$^+$ cation but rarely bound the other metal cations. Thus, the geometry optimization of cation-ligating heterocycles provided highly selective cation receptors.

2.3
Chirality Optimized Receptors

Many kinds of chiral ligands have been developed and applied to use in asymmetric catalysis, enantiomer-selective extraction, chirality sensing, biomimetic modeling, and other aymmetric processes. Since some chiral bio-ligands also exhibit excellent functions in the non-asymmetric biological processes, chirality optimization of the ligand is a promising strategy for designing specific receptors of spherical metal cations [16, 18]. Naturally occurring lasalocid ionophore **14a** is known to mediate biomembrane transport of spherical Na$^+$ cation [19]. This has a series of asymmetric carbons in the acyclic polyether skeleton to promote the pseudo-cyclic metal complexation. Erickson and Still demonstrated that biological lasalocid **14a** exhibited much higher binding constant than non-biological stereoisomers **14b–d** (Fig. 4) [20]. Although it did not target the chiral substrates, the optimization of ligand stereochemistry greatly enhanced the ionophoric functions. Dai, Xu, and Canary prepared chiral ligand **15** (Fig. 4) having a tris(2-pyridylmethyl)amine skeleton, which gave the optimized coordination geometry for specific transition metal cations [21]. These examples strongly suggest that stereo-controlled ligands containing heterocyclic donors can function as a new type of cation-selective receptors.

Kataoka et al. developed a series of N$_3$,O-mixed donor tripode ligands **16** in a stereo-controlled fashion, and characterized their lanthanide cation complexation behaviors [22]. They included two quinoline groups as soft co-ordinating donors and intense chromophores, and also an amide function for the effective binding of lanthanide cations. These formed 1 : 1 and 2 : 1 (tripode : lanthanide cation) complexes with lanthanide nitrates in solution, but the attachment of – CH$_3$ substituents on the tripode skeleton remarkably changed the preferred stoichiometry of the lanthanide complexation. Unsubstituted tripode **16a** formed more stable 1 : 1 complex than 2 : 1 complex with La(NO$_3$)$_3$, while disubstituted tripode **16c** preferred 2 : 1 complex to 1 : 1 complex. The two – CH$_3$ substituents caused severe steric hindrance around the tertiary nitrogen atom and destabilized the 1 : 1 complexation.

Yamada et al. also prepared a series of stereoisomers of tris(2-pyridylmethyl)amines **17** by combining lipase-catalyzed optical resolution and S$_N$2-type replacement reaction [23]. When their log K_1 values for Eu^{3+} complexes were compared, un- and mono-substituted tripodes **17a** and **17b** gave larger log K_1 values than both diastereomers of disubstituted tripodes **17c** (Table 1). Although each tripode had three pyridine and one tertiary nitro-

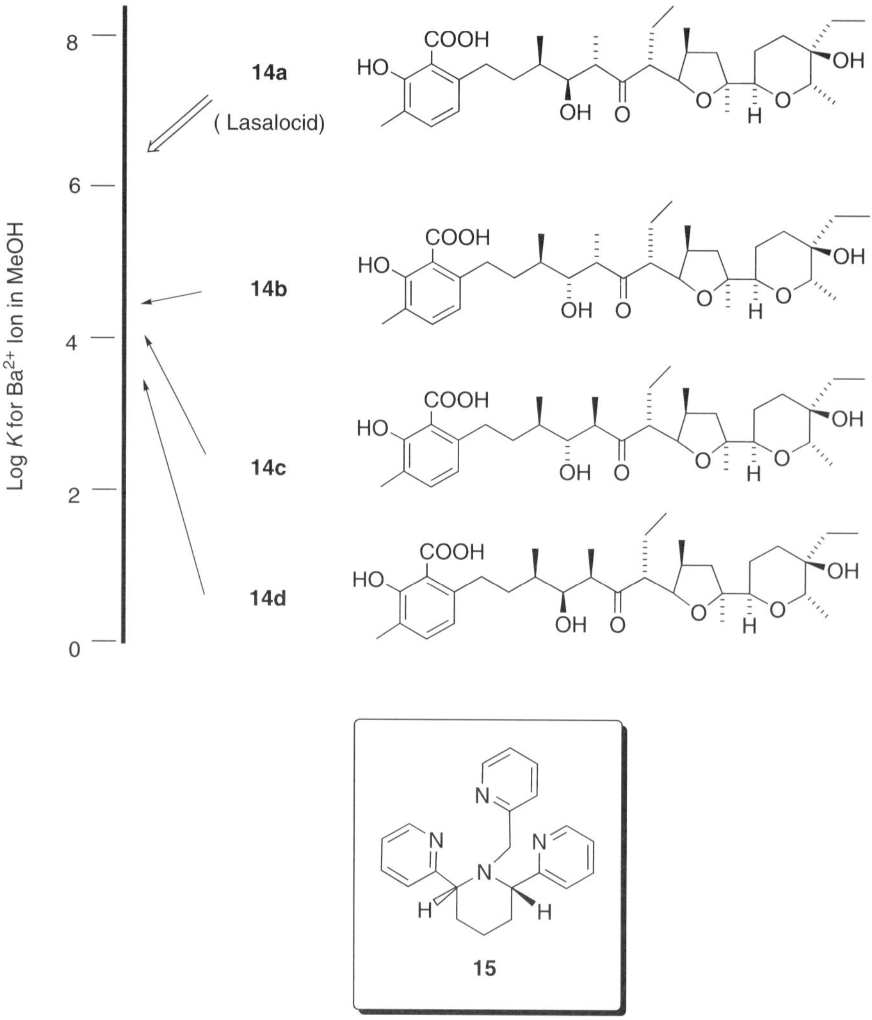

Fig. 4 Cation complexation of stereoisomers of lasalocid **14** and structure of synthetic receptor **15**. Reprinted with permission from [16]. © (2000), American Chemical Society

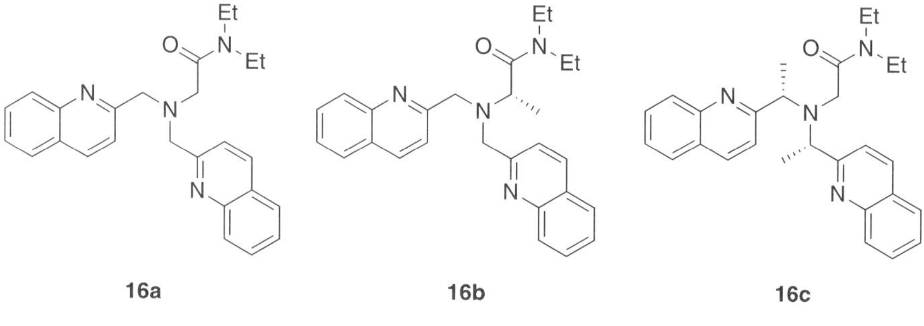

Table 1 Stepwise formation constants for the complexation between tris(2-pyridylmethyl)amine ligands **17** and Eu(CF$_3$SO$_3$)$_3$. Reprinted with permission from [23]. © (2003), American Chemical Society

	17a	(R)-17b	(R,R)-17c	(R,S)-17c
log K_1	7.5 ± 0.4	7.4 ± 0.3	6.9 ± 0.2	6.7 ± 0.3
log K_2	5.2 ± 0.2	5.6 ± 0.2	5.3 ± 0.3	5.5 ± 0.4
log K_3	5.3 ± 0.3	5.4 ± 0.2	5.6 ± 0.1	5.7 ± 0.5

gen atoms as coordination sites, the two introduced – CH$_3$ substituents of disubstituted tripodes **17c** destabilized the 1 : 1 complexation due to the steric problems around the tertiary nitrogen atom. In contrast, the two diastereomers of tripodes **17c** had almost the same log K_1 – K_3 values with Eu^{3+} cation. The stereochemical effect on the lanthanide complexation was not observed among these diastereomers.

2.4
Designed Heterocycles for Supramolecular Cation Recognition

Extensions from cation recognition to supramolecular recognition have recently been done in both natural and synthetic receptor chemistry. Four types of receptor–guest complexes are generally postulated based on the molecular sizes: (1) small receptor–small guest complexes, (2) small receptor–big guest

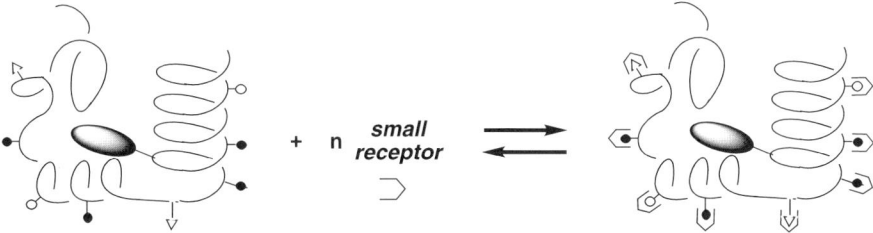

Fig. 5 Small receptor–big protein guest complexation. Reprinted with permission from [144] © (2000), American Chemical Society

complexes, (3) big receptor–small guest complexes, and (4) big receptor–big guest complexes. Proteins are typically recognized by both small and big receptors to offer very sophisticated functions in nature. Although several receptors to activate protein structures and further modify their functions have been found, a chemo-genetic method has received recent attention, in which small receptor molecules formed $n:1$ (receptor : guest protein) complexes to alter the biological protein structures and generated non-biological functions (Fig. 5) [24, 25].

Several crown ether derivatives bound $-NH_3^+$, $-CO_2^-M^+$, or other functional moieties exposed on the protein surface, and formed $n:1$ supramolecular complexes. van Unen et al. [26] and Itoh et al. [27] have demonstrated that the crown ether complexation remarkably enhanced reactivity

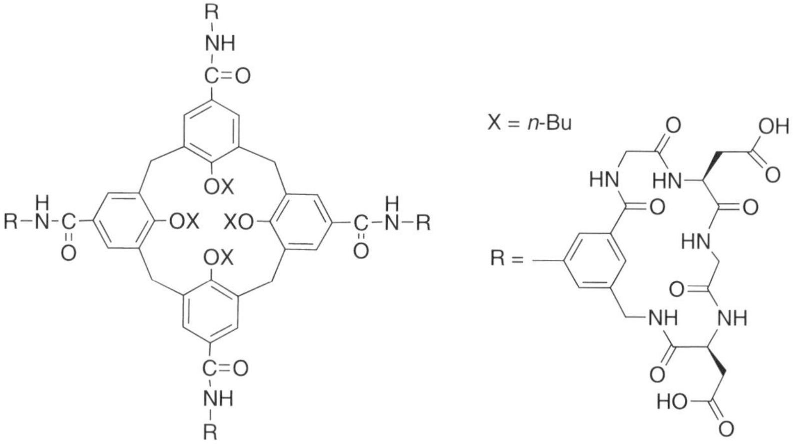

18a

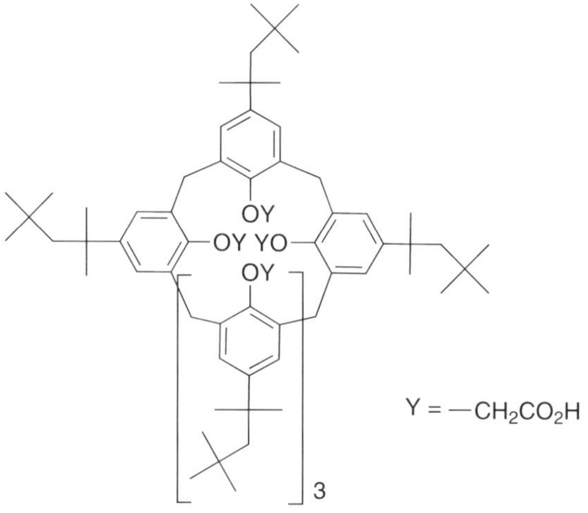

18b

$Y = -CH_2CO_2H$

and enantiomer-selectivity of the hydrolytic enzymes. Cytochrome *c* proteins also behave as big cationic guests, because several protonated lysine moieties locate on the surface. Odell and Earlam had reported earlier that 18-crown-6 and related macrocycles solubilized the water-soluble cytochrome *c* into organic solvents upon supramolecular complexation [28]. Julian and Beauchamp directly observed the supramolecular complexes between 18-crown-6 and cytochrome *c* using the ESI-MS method [29]. Paul et al. further revealed that the cytochrome *c* complexes with various crown ethers had uncommon heme features as synzymes in methanol [30]. Although cytochrome *c* proteins mediate electron transfer processes in mitochondrial respiratory chains and do not work as catalysts in nature, asymmetric oxidation of several sulfoxides occurred at low temperature with crown ether complexes. Hamuro et al. designed calixarene receptor **18a** having polycar-

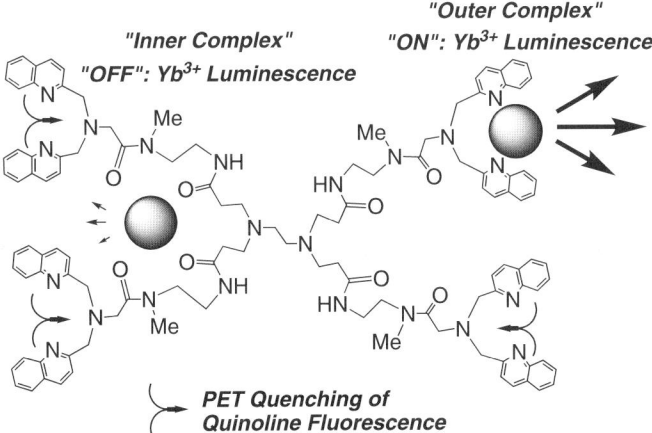

Fig. 6 Dendrimer ligand **19** and Yb^{3+} cation recognition [35]. Reproduced with permission from The Royal Society of Chemistry

boxylic acids to bind the positively charged cytochrome c [31]. Oshima et al. further applied cytochrome c complex with calixarene **18b** in the dye oxidation reaction [32].

As nanoscale big receptors, dendrimer ligands simultaneously incorporate several metal cations in the restricted domains, and provide unique cation binding phenomena [33, 34]. Tsukube et al. constructed a dendrimer ligand that worked as a lanthanide container exhibiting "on-off" switchable luminescence (Fig. 6) [35]. Dendrimer **19** had two different kinds of coordination sites for the lanthanide cations. Each tripode unit on the dendrimer periphery included two quinoline nitrogen, a tertiary nitrogen, and an amide oxygen donor atoms, and formed stable 1 : 1 complexes with $Yb(NO_3)_3$ to give intense luminescence upon irradiation of its quinoline chromophore. The inner polyamidoamine core provided pentadentate coordination for the lanthanide cations, in which the quinoline chromophores stood apart from the bound lanthanide center, and the energy transfer for lanthanide luminescence rarely occurred. The addition of SCN^- anion dramatically altered the coordination environment around the lanthanide centers from inner to outer. Thus, the present dendrimer dynamically switched the lanthanide coordination mode and luminescence profile in response to the external guest anion. Although nano-scaled molecular recognition systems still have many synthetic problems, several dendrimer ligands have successfully been presented for this purpose. Heterocyclic chemistry provides further developments of the nano-scaled cation receptors, molecular machines, functional devices, optical probes, and related supramolecular assembly systems.

3
Anion Recognition with Designed Heterocycles

3.1
Heterocycles for Anion Recognition

Anions are ubiquitous in biological and artificial systems. They have a variety of structural shapes: linear, trigonal planar, tetrahedral, spherical, and others. In addition to inorganic anions, many biomolecules such as nucleic acids, lipids, and ATP exist in the anionic forms. Molecular recognition chemistry focusing on these anionic guests began following the development of cation recognition chemistry with crown ethers and cryptands. Since the anions are much larger in size than the isoelectronic metal cations, large macrocyclic polyammonium cations have been reported to form anion inclusion complexes **20–23** [36–40]. These earlier examples emphasized the significance of the size matching between the receptor cavity and its target anion. However, Coulombic interaction between a receptor and its guest anion is not always effective due to the low charge density of the anions. Hydrogen bonding between a receptor and an anion is also an important interaction in anion recognition, but both interactions are weaker than the metal coordination bond. Several Lewis acids such as metal ions and borates have been incorporated in anion receptors to effectively bind Lewis base types of anions. As found in biological anion recognition systems, a proper arrangement of the multiple interaction sites is required to construct a highly selective receptor for a specific anion. Anion recognition chemistry has been greatly developed in the last few decades, and its progress has been reviewed in many books and review articles [41–43]. We focus here on organic receptors for specific anions, in which heterocycles work effectively as anion binding sites.

Pyridine, imidazole, pyrrole, and other nitrogen-containing heterocycles have been used as useful building blocks of anion receptors. They have several advantages as anion binding sites: (1) the heteroaromatic increases receptor rigidity to exhibit high anion selectivity, (2) the N-alkylated heterocycle has positive charges delocalized to work as a soft binding site, and (3) the hydrogen bonding is available at the designed position. When N-alkylated pyridines and imidazoles were employed as quaternized heterocycles, the proton connected to the carbon atom next to the quaternary nitrogen atom formed a C–H···anion hydrogen bond with the guest anion (Fig. 7). Bryantsev and Hay recently pointed out the importance of such hydrogen bondings in the anion binding process [44]. Since anion–π interactions with electron deficient aromatic compounds were also recognized as a new anion binding motif [45], the N-alkylated heterocycles can operate as effective binding sites for electrostatic and hydrogen bonding interactions. Pyrroles are alternative building blocks for anion receptors. Their N–H protons are relatively acidic ($pK_a = 17$) and bind several anions. Amides, ureas, and guanidines also have

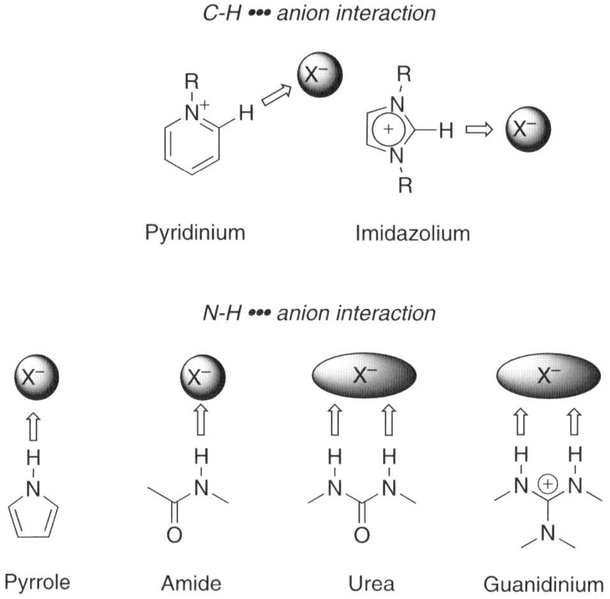

C-H ••• anion interaction

Pyridinium **Imidazolium**

N-H ••• anion interaction

Pyrrole **Amide** **Urea** **Guanidinium**

Fig. 7 Hydrogen bonding interactions for anion binding

N – H bonds available for hydrogen bondings, and were incorporated in several receptor molecules [46]. Since these hydrogen bondings are fairly stable in highly polar media, the designed anion receptors had cooperative multipoint interactions with the specific guest anions. Thus, multiple hydrogen bondings offer complementary pairings with bidentate anions and simultaneous bonding from several spatially separated sites (Fig. 8). Organization of the multipoint hydrogen bonding sites in the macrocyclic skeleton provides

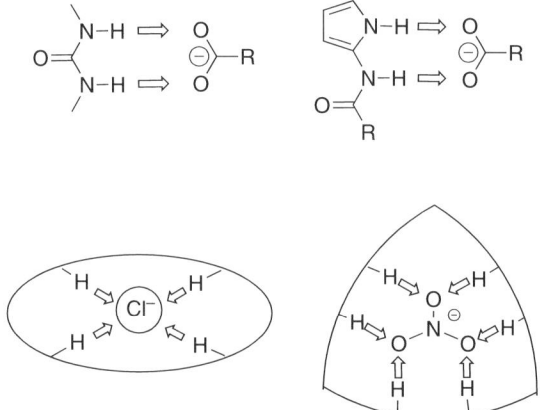

Fig. 8 Multiple hydrogen bonds toward anions

high anion selectivity based on matching the size, direction, and shape. In the following sections, recent examples of designed heterocycles as anion receptors and their applications are summarized.

3.2
Cationic Heterocyclic Receptors

Multi-charged receptors offer effective anion recognition phenomena in water, because they can strongly attract negatively charged species by the Coulombic interaction [47, 48]. Cationic cyclophane **24**, a quaternary tetraammonium macrocycle containing rigid benzene rings, had been introduced by Tabushi et al. [49]. This showed $N^+ \cdots$ oxoanion interaction to effectively promote hydrolytic catalysis of aromatic esters, in which hydrophobic interaction between aromatic moieties further occurred [50].

24

Pyridine and imidazole work as bases in aqueous solutions, but the protonated heterocycles do not operate well as anion receptors due to their weak basicity (pK_a(pyridine) = 5.25, and pK_a(imidazole) = 6.82). In contrast, N-alkylated pyridinium and imidazolium cations function as effective binding sites, of which charged states are not dependent on the pH of the solution. Cramer et al. prepared cyclic pyrimidinium tetramer **25a** [51] and hexamer **25b** [52] by degradation of thiamine (vitamin B_1) in methanol (Fig. 9). Nucleophilic displacement of the thiazole unit gave oligomers of pyrimidinium salts. Thiamine chloride and nitrate also formed cyclic tetramers, which accommodated the anions via hydrogen bonding from the aromatic C – H bonds. Since larger inorganic anions like $[Ba(NO_3)_6]^{4-}$ promoted the formation of cyclic hexamers, the anions were thought to work as template ions during the cyclization reactions [53, 54]. Unfortunately, they had low solubilities in water and methanol, and their anion receptor functions were not characterized.

Shinoda et al. reported a similar cyclization reaction with 3-bromomethylpyridine to give cyclic pyridinium tetramer **26** (Fig. 10) [55]. The protons of the cyclic tetramer had acidic nature, and were rapidly deuterated in D_2O so-

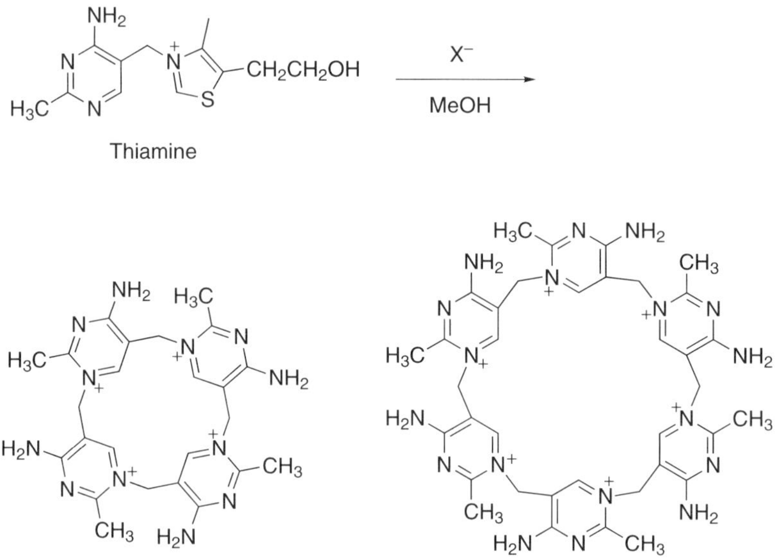

Fig. 9 Synthesis of cyclic pyrimidinium tetramer **25a** and hexamer **25b**

lution. Crystal structure revealed that two of the four Br⁻ anions located above and below the 1,2-alternative macrocycle, and formed hydrogen bonding with the aromatic protons. This tetramer bound tricarboxylate anions more strongly than mono- and di-carboxylates in water. Due to the large electrostatic interactions, this macrocycle showed large 1 : 1 binding constants ($\log K = 4 \sim 5$) for tricarboxylate anions with highly negative charge densities.

A series of calixarene-type cyclic oligomers were systematically prepared. Cabildo et al. synthesized dicationic macrocycles **27** from α,α'-dibromo-p-xylene with pyrazole and imidazole [56]. The energy barrier of the ring flipping interconversion was estimated in solution ($\Delta G^{\ddagger} = 17$ kcal mol^{-1} for

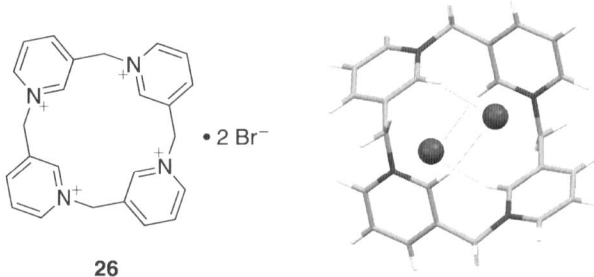

Fig. 10 Cyclic pyridinium tetramer **26** and its Br⁻ complex structure

27a). Alcalde et al. determined the crystal structure of dicationic imidazolium macrocycle **28** in which two Cl⁻ anions formed two hydrogen bonds with the C–H protons of the receptor (Fig. 11) [57]. The quantitative investigations revealed that the ring closure reaction was promoted by Cl⁻ complexation with the monocationic linear precursor [58]. Recently, Chellappan et al. reported the crystal structure of calix[4]imidazolium[2]pyridine **29**–F⁻ complex (Fig. 11), where the imidazolium C-H protons participated in hydrogen bonding with the F⁻ anion [59].

When several cationic moieties were organized in the conformationally regulated non-macrocycle to match the shape of a guest, the electrostatic interactions operated in the specific binding of the anionic guests. 1,3,5-Trisubstituted benzene provided an effective scaffold to arrange three func-

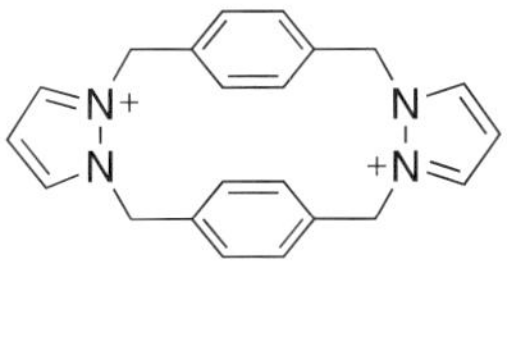

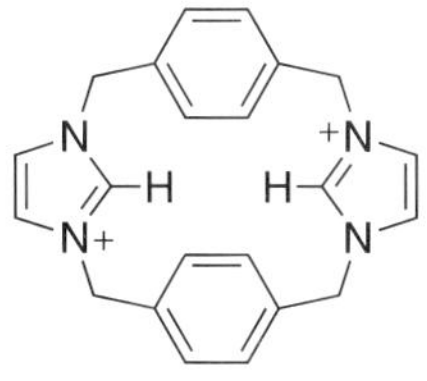

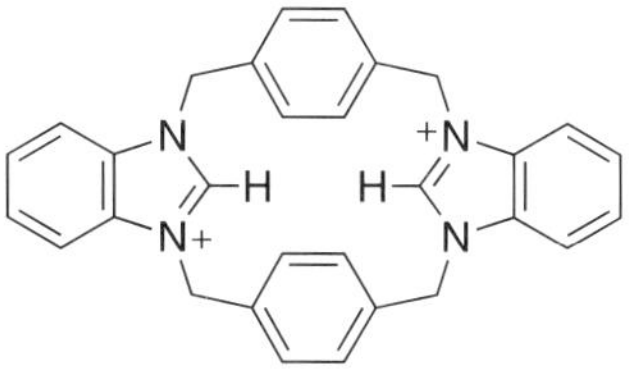

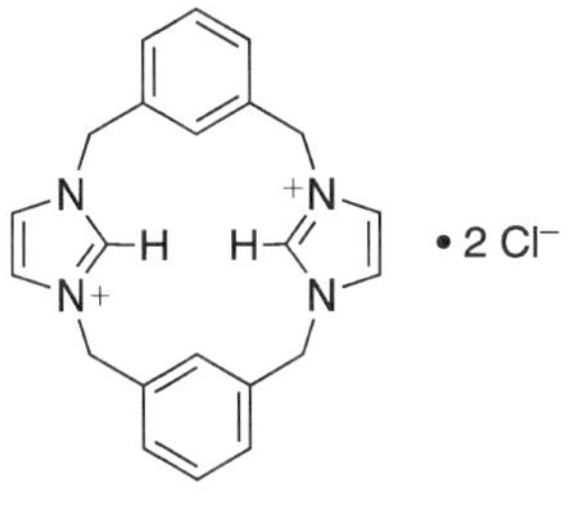

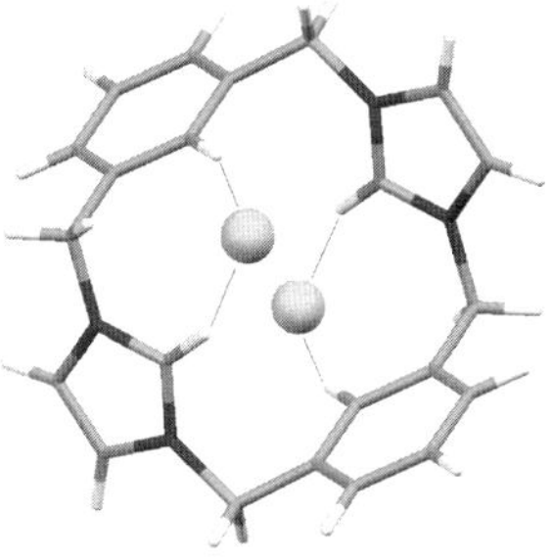

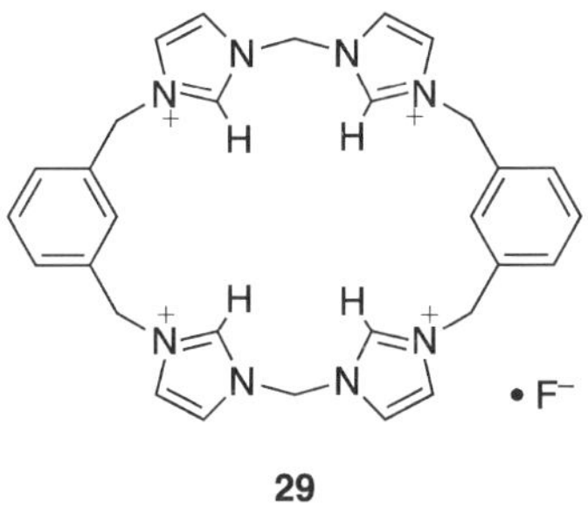

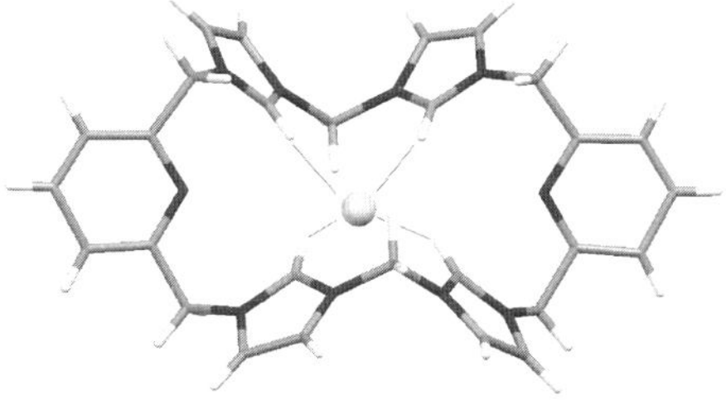

Fig. 11 Cyclic imidazolium receptors **28** and **29**, and their crystal structures

30 **31**

tional groups in a C_3 symmetric fashion. Sato et al. reported that tripodal receptor **30** having three imidazolium sidearms formed halide anion complexes through $C-H\cdots X^-$ interactions [60]. Ihm et al. reported that tripodal receptor **31** exhibited Cl^- and $H_2PO_4^-$ anion selectivity [61]. Yoon et al. recently applied other imidazolium-containing anion receptors in fluorescence sensing [62]. Abouderbala et al. prepared a tripyridinium receptor **32** of which *ortho*-$C-H$ protons offered cooperative $C-H\cdots Br^-$ interaction in CH_3CN solution [63]. The X-ray crystal structure of the complex indicated that one Br^- anion located within the cavity and was supported by three 3-aminopyridinium arms (Fig. 12) [64]. Ilioudis et al. synthesized the macrobicyclic receptor of halide anions based on the 1,3,5-trisubstituted benzene. In its F^- complex **33**, the anion was completely encapsulated into the cage of the protonated receptor supported by $N^+-H\cdots F^-$ and $C-H\cdots F^-$ types of hydrogen bondings (Fig. 12) [65].

These cationic *N*-alkylated heterocycles attracted the specific anions by electrostatic interactions even in highly polar solvents. $C-H\cdots X^-$ hydrogen bonding also played an important role in the cooperative binding of the anions. The use of appropriate molecular scaffolds particularly enabled three-dimensional arrangements of binding sites to provide cooperative binding of the specific anions.

3.3
Neutral Heterocyclic Receptors

As acidic $N-H$ protons are good hydrogen bond donors for basic guest anions, amide, urea, guanidinium, and pyrrole derivatives have been incorporated in macrocyclic receptors of various dimensions. Several neutral receptors have successfully been designed to match the size and the shape of their target anions. Since the hydrogen bonding interactions between receptors and guests did not usually work in protic media, these anion receptors have been characterized in aprotic media such as DMSO, CH_3CN, and $CHCl_3$.

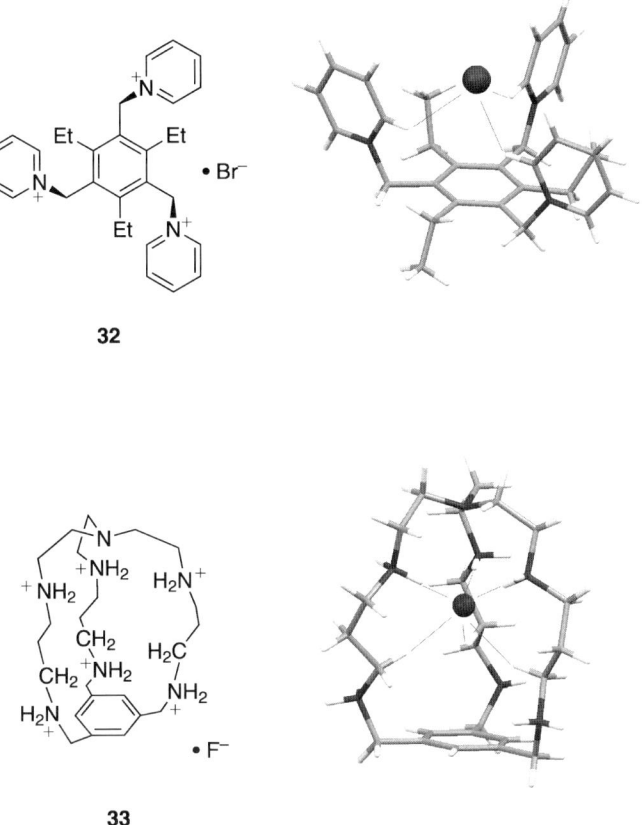

Fig. 12 Crystal structures of anion complexes with **32** and **33**

Gale et al. recently reviewed designs of anion receptors [66–69], while amide-based macrocyclic receptors were detailed by Kang et al. [70, 71].

Among the various heterocycles, porphyrins have most often been characterized, which have two pyrrolic N–H groups inside the macrocyclic π-conjugated system. Although these strongly bound various metal cations, they did not work as effective anion receptors. Due to the rigid aromatic framework and the limited space inside the porphyrin core, anions cannot be incorporated into the center cavity. Sessler et al. developed a series of expanded porphyrins in which more than four pyrrole rings were included in conjugated macrocycles. Sapphyrin, a pentapyrrolic expanded porphyrin, typically formed the F$^-$ inclusion complex **34** upon di-protonation [72], in which the F$^-$ anion was located at the center of the flat di-protonated sapphyrin and bound by five N–H···F$^-$ hydrogen bondings. Since the larger Cl$^-$ anion was incompletely embedded in this macrocyclic plane, the ring size of the expanded porphyrin determined the anion selectivity [73].

34

Gale et al. presented calix[4]pyrrole **35** as an effective anion binding agent [74]. The pyrrole rings of the calixpyrrole were connected at their α-positions by quaternary carbon atoms. As observed in the crystal structures of both Cl⁻ and F⁻ complexes (Fig. 13), each pyrrole N–H group worked independently as a hydrogen bond donor for the anion. Sessler et al. prepared larger macrocyclic receptors **36–38** containing pyrrole and 2,2′-bipyrrole units (Fig. 14) [75–77]. The replacement of a single carbon atom of **35** with benzene ring greatly enhanced the anion binding strength through hydrogen bondings especially for the small Cl⁻ and NO_3^- anions (see **38**).

A variety of hydrogen donor groups were combined with pyrrole to allow cooperative bindings with anions. Gale et al. reported that acyclic amidopyrroles **39** worked as anion binding agents [78]. The crystal structures of their anion complexes revealed complementary hydrogen bondings with carboxylate anions [79]. Chmielewski et al. employed carbazole to prepare a rigid receptor **40**, of which N–H groups cooperatively bound anions [80]. Maeda et al. connected two pyrrole rings to a β-diketone unit [81]. The complexation with BF_3 gave anion receptor **41**, which showed high binding constant with acetate anion in CH_2Cl_2 via hydrogen bondings with the α-C–H proton [82].

Chang et al. developed indole-based macrocycles **42** that had four indole N–H groups directed towards the inside of the large cavity (Fig. 15) [83]. The

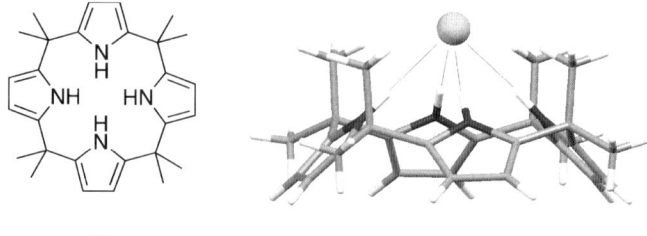

35

Fig. 13 Calix[4]pyrrole **35** and crystal structure of its F⁻ complex

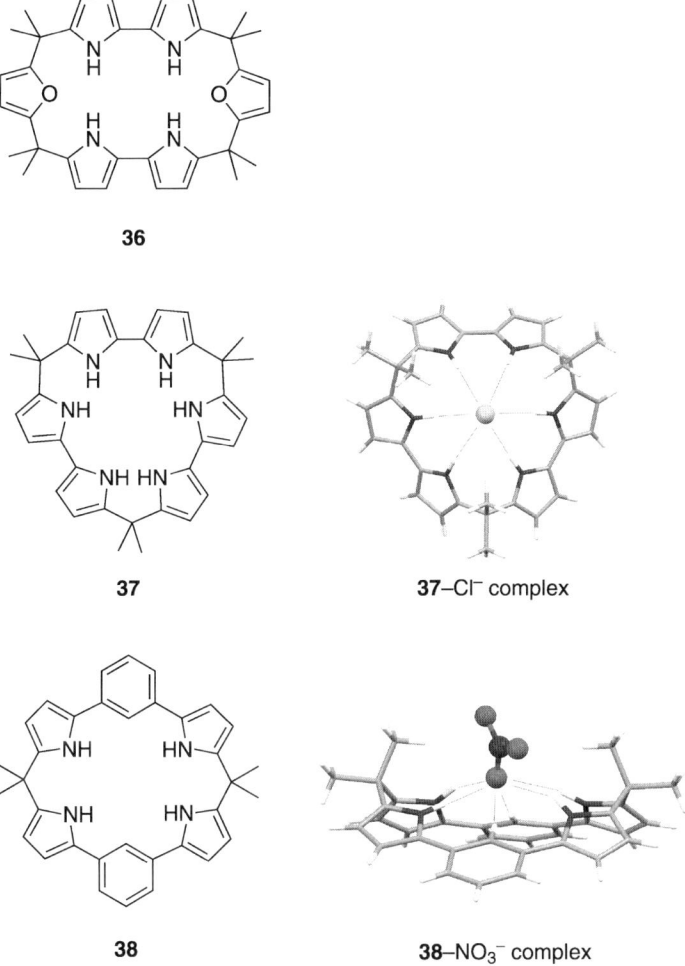

Fig. 14 Pyrrole-containing macrocycles **36–38** and crystal structures of their anion complexes

association constants decreased in the order of $F^- > Cl^- > Br^- > I^-$. Because the anion-induced ^{1}H NMR signal shifts of the N – H protons were dependent on the type of coordinating anion, and since different anion guests exhibited separated NMR signals at the same time, receptor **42** acted as an NMR probe to detect the anions. Katayev et al. prepared amide-imine macrocycles **43**, of which pyrroles effectively bound larger oxoanions such as acetate, HSO_4^-, and $H_2PO_4^-$ anions [84]. Crystal structure analysis and DFT calculation revealed that the flexible macrocycle adopted conformations suitable for hydrogen bondings to these large oxoanions (Fig. 15).

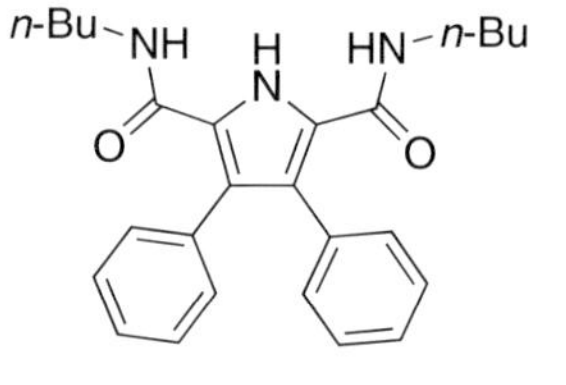

Fig. 15 Macrocycles 42 and 43 with large cavities for anion binding

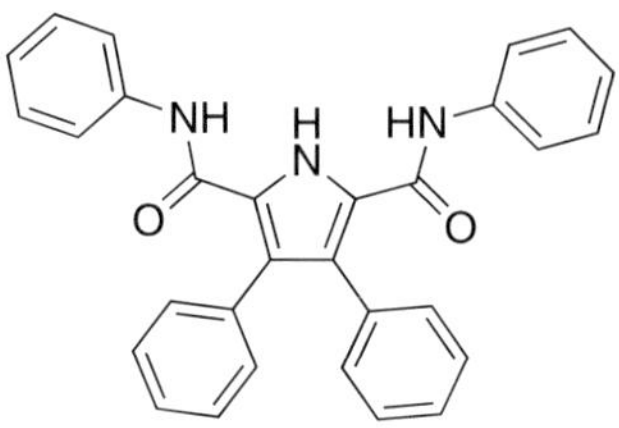

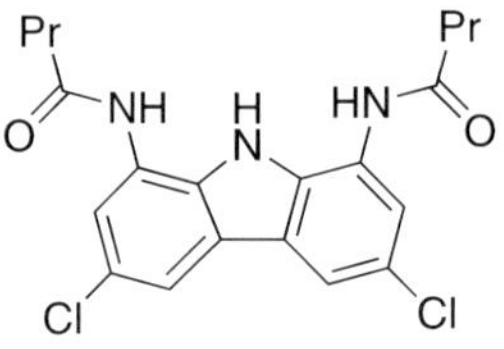

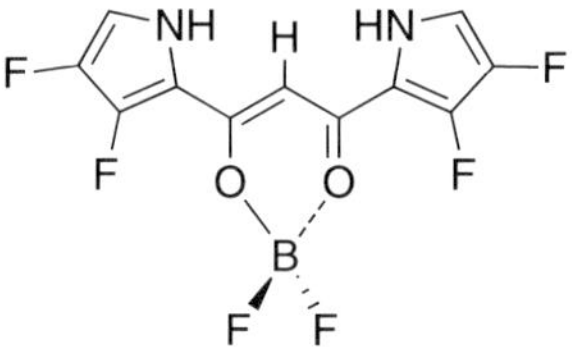

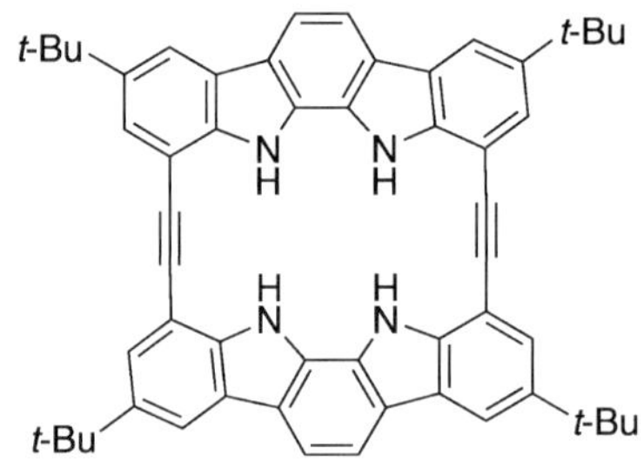

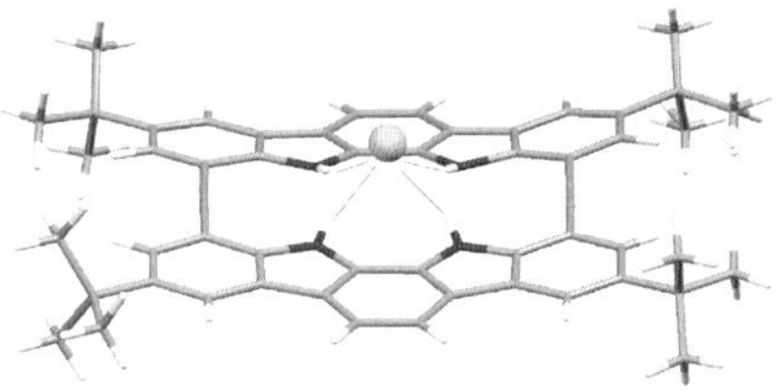

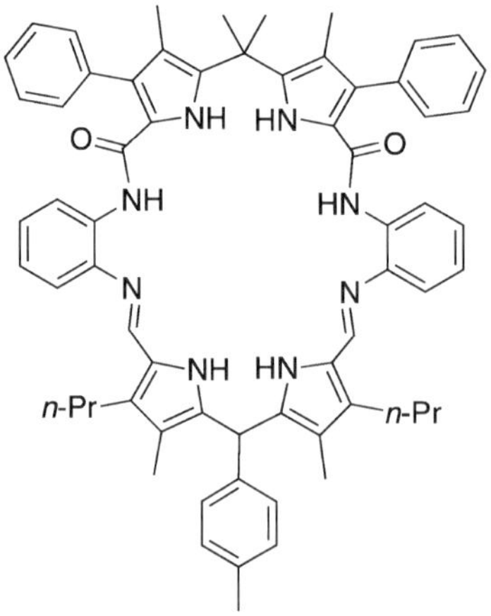

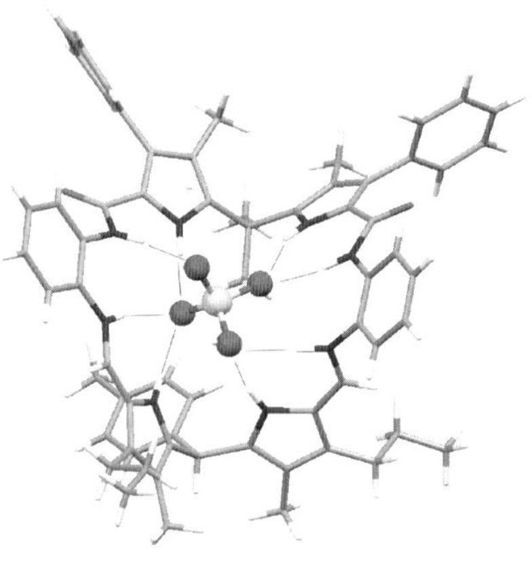

3.4
Anion Detection with Designed Heterocycles

When a chromophoric group of the receptor is electronically perturbed by the anion binding, anion can be detected by monitoring the color change of the receptor [85, 86]. Black et al. reported that dipyrrolylquinoxaline **44** worked as a chromogenic signaling agent for F^- anion [87]. This receptor showed high selectivity for F^- over Cl^- and $H_2PO_4^-$ anions via cooperative action of two pyrrole subunits. The introduction of a nitro group in the aromatic moiety enhanced the anion binding constant with the F^- anion, and also caused a vivid color change from yellow to purple. Nishiyabu and Anzenbacher derived chromogenic sensors **45** from calix[4]pyrrole **35**, which strongly bound F^-, acetate, and $HP_2O_7^{3-}$ anions and caused drastic color changes [88]. Since these showed much higher affinity for acetate anion than Cl^- and HPO_4^{2-} anions, they were used to detect carboxylates under physiological conditions. Evans et al. prepared pyrrolylamideurea **46a** as a colorimetric anion receptor [89]. When F^- or acetate anion was added, the hydrogen bonding interactions caused color changes of receptor **46a** to dark yellow. When thiourea

46b was used, the addition of the anion gave a yellow-to-red color change due to the deprotonation of the thiourea.

Indicator-displacement assay [90] is an effective method that allows optical anion detection with optically inactive receptor. If a colorimetric or

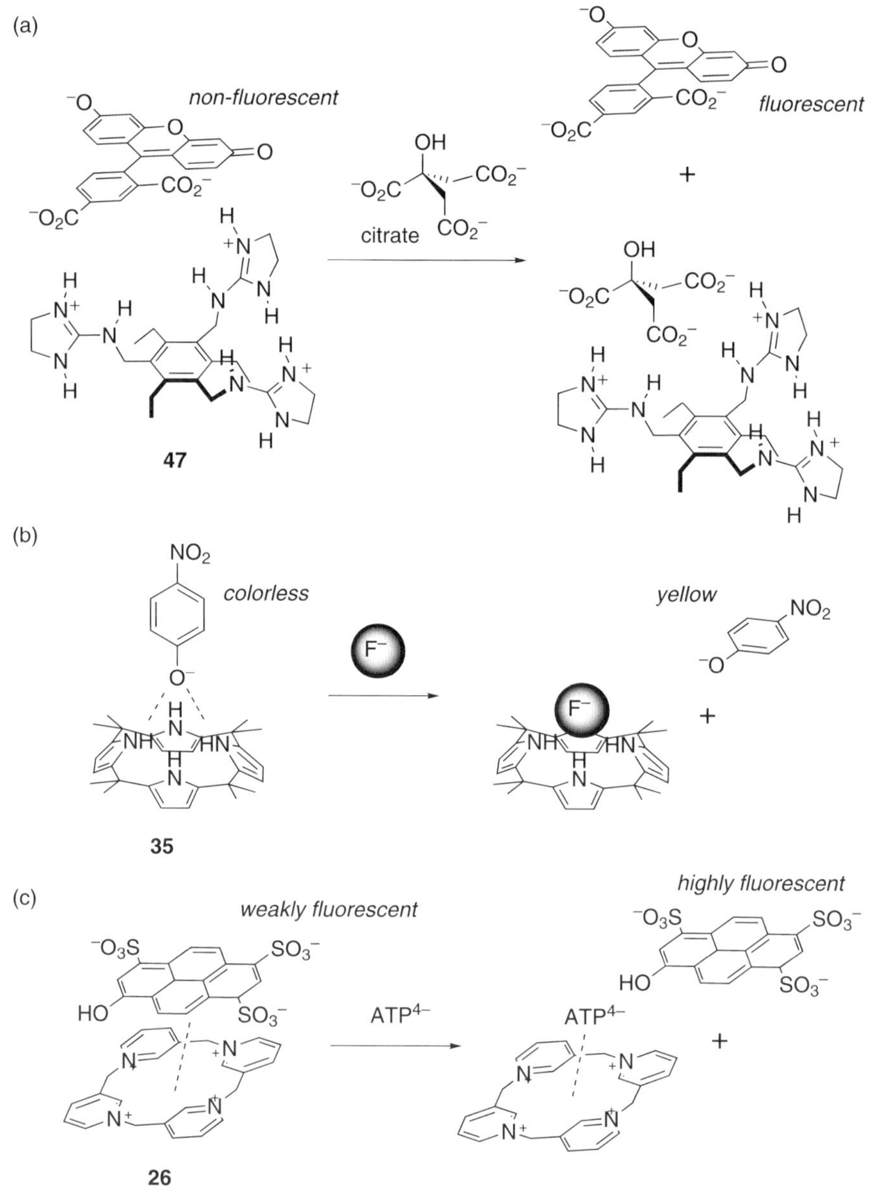

Fig. 16 Displacement assay of non-chromophoric guest anions

fluorescent indicator complexed with the receptor before the addition of analyte, a specific analyte displaces the indicator and causes an optical response. Although various anion sensing molecules have been prepared by covalent attachment of indicator molecules, several non-chromophoric receptors were applicable in the displacement assay (Fig. 16).

Metzger and Anslyn applied this method in the citrate anion detection using a tricationic tripode **47** [91], in which 5-carboxyfluorescein interacted with guanidinium groups as the indicator. They also attached boronic acid groups with the receptor that well sensed tartarate [92] and 3,4,5-trihydroxybenzoate anions [93]. Gale et al. enabled the optical detection of F^- anion using a combination of p-nitrophenolate indicator and calix[4]pyrrole receptor **35**. The yellow color of the free indicator appeared only when it was released from calix[4]pyrrole by the more strongly bound F^- anion [94]. Tetracationic cyclic pyridinium receptor **26** was combined with 8-hydroxy-1,3,6-pyrenetrisulfonate to allow fluorescence detection of ATP anion in water [95]. Although the indicator bound to the cationic receptor showed weak fluorescence, ATP anion displaced the indicator to cause the enhanced fluorescence. Since ADP and AMP anions were not bound in aqueous solution, this worked as a selective receptor for polyanionic species.

As described above, multiple hydrogen bonding interactions were significantly involved in most anion recognition systems as well as complementary electrostatic interactions. Pyrrolic N–H group particularly worked as an efficient hydrogen bonding donor, while the significance of C–H··· anion interaction was recently recognized. Since these hydrogen bonding interactions caused large changes in NMR chemical shifts and electronic states of the chromophores, several anion detection systems have been constructed using the designed heterocycles.

4
Anion Recognition with Designed Heterocycle–Lanthanide Complexes

When an anion-selective receptor is designed using metal complex, the shape, geometry, charge, and hydrophobicity of its target must be considered. Various metal complexes have already been developed as anion receptors, in which electrostatic interactions, hydrogen bonds, and metal coordinations were effectively combined [96]. Recent attention has particularly focused on luminescent lanthanide complexes applicable in biological systems [97]. Eu^{3+} and Tb^{3+} complexes are the promising families which have practical potentials as luminescence anion-sensing probes [98–100]. This section covers the anion recognition with receptors composed of $4f$-metals and heterocyclic ligands. Although a large number of target anions of natural origin are water-soluble, some heterocycle–lanthanide complexes allowed their sensing at a practical level. Their synthetic strategies and sensing characteristics are discussed.

4.1
Heterocycle–Lanthanide Complexes for Anion Recognition and Sensing

The trivalent lanthanide cations possess characteristic $4f$ open-shell configurations and exhibit interesting chemical and physical properties. Since these have large ionic radii ranging between 0.89 Å and 1.16 Å in the octa-coordinated complexes, most of them prefer high coordination numbers (8–10). When the coordination number of the employed ligand does not match, the remaining coordination sites are occupied by counter-anions and/or solvent molecules [98, 99]. In such a case, the lanthanide complex can form adducts with external guest anions, which are called ternary, mixed, or highly coordinated complexes. Since some lanthanide cations have characteristic Lewis acidities, light emitting properties, and magnetic functions, a proper combination of the lanthanide center with a designed heterocyclic ligand allows precise anion recognition and highly selective sensing. Several anion receptors of this type have been developed which had one or more free sites within the first coordination sphere of the lanthanide center for incoming guests. Although some heteroaromatics serve as photoantennae and coordination sites, the sophisticated ligand for a lanthanide-based chemosensor should contain (1) a chromophore or a signaling group, (2) adequate coordination sites close to the saturation, and (3) high stability and solubility of the incompletely saturated complex in solution.

Some kinds of heterocycle–lanthanide complexes are known as CD probes for the efficient chirality detection of target guests [99]. The asymmetric arrangement of the chromophoric heterocycles around the lanthanide center was induced by highly coordinated complexation with chiral guests. The sign of the observed CD signals at the ligand chromophore bands was dependent upon the chirality of the guest [100]. Some Yb^{3+} complexes worked as near-IR CD probes for chiral anions in the solution state [101], though CPL spectra were often observed with Eu^{3+} and Tb^{3+} complexes [102]. NMR method is a valuable alternative for the efficient chirality detection of optically active anions [100, 103]. A series of chiral lanthanide complexes are commercially available as chiral shift reagents for the enantiomeric purity determination. Most often the enantiopure form of the lanthanide complex was used for resolving the enantiomeric pair of a given substrate [100, 104–106], whereas only a few examples of racemic lanthanide complexes discriminated enantiomers of the guests [107–109].

4.2
Luminescent Lanthanide Complexes

Luminescence analysis is a promising tool in analytical chemistry, biochemistry, and cellular biology [110, 111], because its simplicity and high sensitivity offer the sensing and detection of chemical traces [112, 113]. While most transition metal complexes act as quenchers, Eu^{3+}, Tb^{3+}, and other lan-

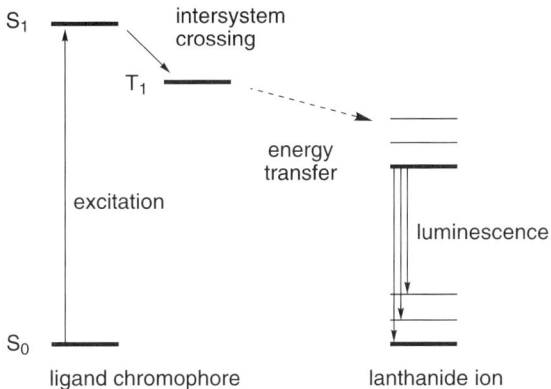

Fig. 17 Schematic representation of "antenna-effect" principle in lanthanide luminescence

thanide complexes display long-lived, line-shaped, and position-fixed emission signals [114], which are sensitive to ligand characters and coordination environments [98, 100]. Eu^{3+} and Tb^{3+} cations exhibit red and green luminescence with millisecond lifetimes, while Nd^{3+} and Yb^{3+} cations have emission with microsecond lifetimes in the near-IR region. The former are applicable for detection by the naked-eye, while the latter can have in vivo applications [115–119]. Since the lifetimes of these lanthanide complexes are much longer than those of most organic fluorophores (ca. 10 ns), several Eu^{3+} and Tb^{3+} complexes have been employed as labeling reagents in time-delayed fluorescence assays [120–123]. Their luminescent lanthanide centers were completely shielded from the external environments. Thus, they exhibited intense emission signals, but showed modest anion-responsiveness.

Owing to the Laporte-forbidden $f-f$ transitions and low absorption coefficients of the lanthanide cations, excitation of the ligand chromophore is usually required, followed by an energy transfer process (Fig. 17) [124]. The excited state of the lanthanide center is further vulnerable to quenching by non-radiative decay processes such as O – H vibration. Since the energy transfer efficiency is significantly affected by the highly coordinated complexation with external guests, several luminescent lanthanide complexes operated well as anion receptors even in aqueous media [98, 100].

4.3
Lanthanide Complexes for Luminescence Sensing

4.3.1
Anion Sensing in Organic Media

Several types of heterocyclic ligands were reported to form luminescent lanthanide complexes exhibiting anion-selective responses in non-aqueous media. Montalti et al. combined a hard phosphine oxide fragment with two soft bipyridines to give acyclic hybrid ligand **48** [125]. This formed stable complexes with Eu^{3+} and Tb^{3+} cations, in which two water molecules directly coordinated with the lanthanide centers [126]. Addition of NO_3^- anion greatly increased the luminescence intensity of both Eu^{3+} and Tb^{3+} complexes, while F^-, Cl^-, or acetate anion induced less pronounced changes. The displacement of the water molecules by one NO_3^- anion led to a ternary complex exhibiting the enhanced luminescence intensity (A in Fig. 18). Addition of one more equivalent of NO_3^- anion dissociated one bipyridine, and produced another type of ternary complex (B in Fig. 18). Finally, the coordination of a third NO_3^- anion decreased the luminescence intensity due to the release of the second bipyridine arm (see C in Fig. 18).

Yamada et al. applied Eu^{3+} complex with **17b** in anion sensing upon excitation of pyridine antenna in acetonitrile [127]. Among F^-, Br^-, Cl^-, I^-, SCN^-, NO_3^-, ClO_4^-, acetate, HSO_4^-, and $H_2PO_4^-$ anions, only NO_3^- anion gave relevant emission enhancement of the Eu^{3+} complex (ca. 5-fold). On the contrary, Cl^- and acetate anions offered modest enhancements, and the other tested

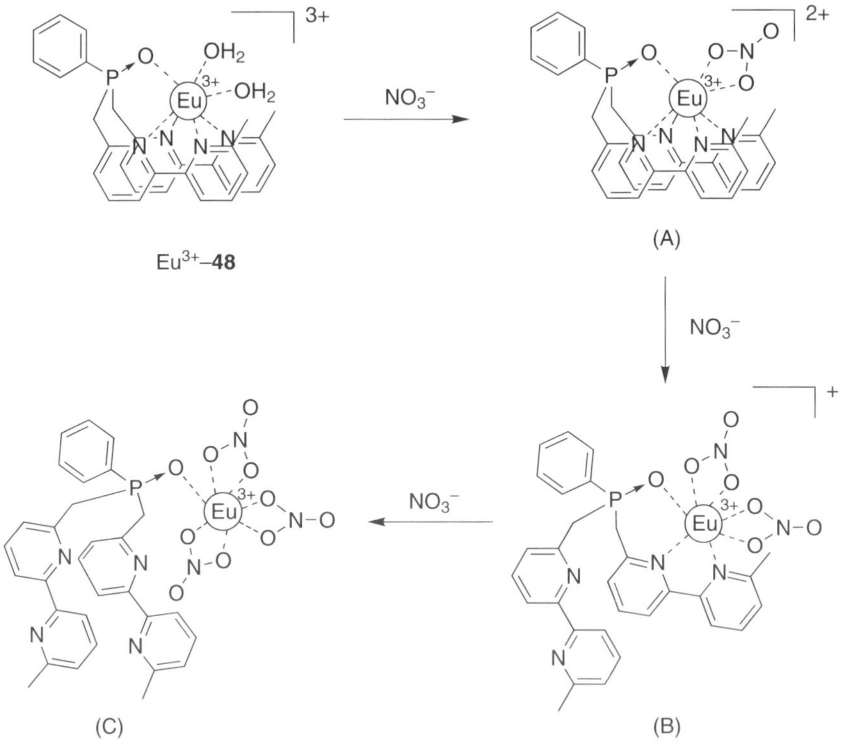

Fig. 18 Proposed mechanism for the stepwise coordination of NO_3^- anion to Eu^{3+}–**48** complex. Reprinted with permission from [126]. © (2002), American Chemical Society

48

anions caused no significant change. When the Eu^{3+} complex with achiral tripode **17a** was employed, the same trend of selectivity was found with lower sensitivity.

Kataoka et al. reported the N$_3$,O-mixed donor tripodes **16** bearing two quinoline chromophores [22]. In acetonitrile, tripode **16a** formed luminescent 1:1 (ligand:lanthanide) complexes with both EuCl$_3$ and EuNO$_3$, where the tripode provided the cooperative coordination of soft quinoline nitrogen, apical nitrogen, and hard amide oxygen atoms to the lanthanide center. In contrast, **16c** formed a luminescent 1:1 complex with EuCl$_3$, but a 2:1 non-luminescent complex with EuNO$_3$. The relative luminescence intensity at 616 nm was monitored after the addition of F$^-$, Br$^-$, Cl$^-$, I$^-$, SCN$^-$, NO$_3^-$, ClO$_4^-$, acetate, HSO$_4^-$, and H$_2$PO$_4^-$ anions. The disubstituted tripode **16c**–Eu(CF$_3$SO$_3$)$_3$ complex exhibited a high selectivity for Cl$^-$ anion with a 63-fold luminescence enhancement, though the unsubstituted tripode **16a** showed similar enhancements for Cl$^-$ and NO$_3^-$ anions.

Best and Anslyn presented hybrid-type receptor **49** (Fig. 19), in which a luminescent Eu^{3+} complex and two quaternary ammonium units were

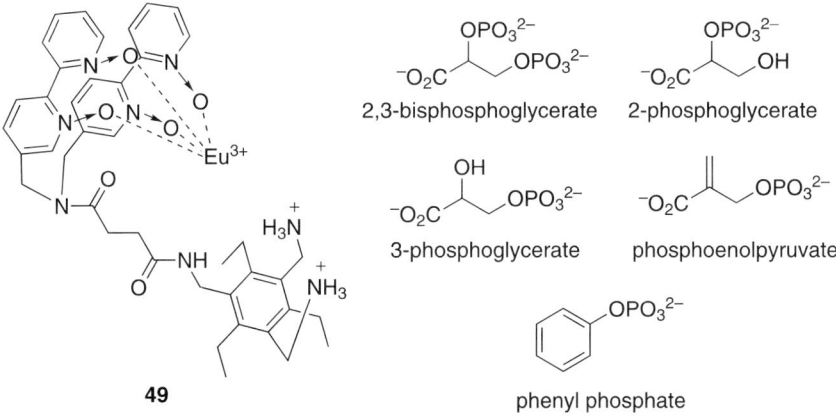

Fig. 19 Luminescent Eu^{3+}–**49** complex for sensing of phosphate anions

combined [128]. In a methanol–acetonitrile (1 : 1) solution, addition of 2-phosphoglycerate, 3-phosphoglycerate, phosphoenolpyruvate, 2,3-bisphosphoglycerate, or phenyl phosphate anion decreased the Eu^{3+} luminescence. The stability constant of the ternary species with 2,3-bisphosphoglycerate anion was three times greater than that with phenyl phosphate anion, indicating that 2,3-bisphosphoglycerate anion was bound to both the Eu^{3+} ion and the two ammonium subunits, whereas phenyl phosphate anion only coordinated to the Eu^{3+} center.

4.3.2
Anion Sensing in Aqueous Media

Dickins and coworkers have developed a series of cyclen–lanthanide complexes as luminescent anion receptors working in aqueous solutions. Cyclens **50** functionalized with three coordinative pendant arms were confirmed to provide 7-coordination to the Eu^{3+} and Tb^{3+} cations, while two sites were available for external anions [129, 130]. Displacement of the two coordinated water molecules by the guest anions rapidly occurred. In the pH range of 5.5–6.5, these complexes exhibited changes in the luminescent properties in response to acetate, F^-, SO_4^{2-}, citrate, lactate, or malonate anion, while Cl^-, Br^-, I^-, and NO_3^- anions gave no significant response.

50a: R = H
50b: R = Me

Mameri et al. developed acyclic ligand **51** containing bipyridine carboxylic moieties, which gave high stability and hydrophilicity of the lanthanide complex [131]. This formed luminescent 1 : 1 complexes with Eu^{3+} and Tb^{3+} cations, where two water molecules located in the first coordination sphere of the lanthanide centers. The efficient ligand-to-metal energy transfer was ensured by the bipyridine photoantenna. Upon addition of ATP anion, the Eu^{3+} luminescence intensity decreased to 20% of its initial value. Since the luminescence lifetime increased from 0.28 to 0.58 and 0.65 ms with the addition of 10 and 20 equivalents of ATP anion, the two bound water molecules were replaced by the external ATP anion. In contrast, ADP, AMP, and NO_3^- anions did not induce any significant change. The ternary complex of Eu^{3+}–**51** with ATP anion was detected by ESI-MS and ^{31}P NMR, and also supported by quantum mechanical calculations [132].

51

52

Ziessel et al. developed ligand **52**, in which two soft 6-carboxy-2,2′-bipyridine arms were directly tethered to a hard phenylphosphine oxide [133]. Its 1 : 1 complexes with Eu^{3+} and Tb^{3+} cations exhibited characteristic luminescence at neutral pH. Addition of hydrogen phosphate, ADP, and ATP anions increased the luminescence intensities and lifetimes of both Eu^{3+} and Tb^{3+} complexes, while AMP and NO_3^- anions induced no obvious change. ATP-induced luminescence enhancement was assigned to a diminution of the non-radiative deactivation processes due to the bound water molecules.

Magennis et al. prepared two azacrown ether derivatives **53** and **54**, which formed 1 : 1 complexes with Eu^{3+} and Tb^{3+} cations [134]. Addition of picolinate anion to a solution of Eu^{3+}–**53** or Tb^{3+}–**53** gave 250- or 170-fold enhancement in the emission signals. Eu^{3+}–**54** exhibited similar augmentations upon the addition of picolinate anion (120-fold). Since these crown ethers were non-chromophoric, the coordinated aromatic carboxylate guests worked as antennae "switching on" the lanthanide luminescence.

Gunnlaugsson et al. isolated the luminescent diaqua Eu^{3+} and Tb^{3+} complexes with cyclens **55** [135]. The X-ray analysis revealed that cyclen **55b** worked as a heptadentate ligand. These non-chromophoric complexes were "photophysically silent" upon excitation at 300 nm, but their Tb^{3+} complexes displayed characteristic emissions via highly coordinated complexation with chromophoric salicylate anion. The excited coordinated salicylate anion sensitized the lanthanide center, and luminescence enhancement factors of ca. 680 and 220 were observed with the two Tb^{3+} complexes. Since acetylsalicylate anion gave no sensitization, both complexes worked as effective chemosensors for salicylate anion under physiological conditions.

Leonard et al. prepared a luminescent ternary complex from ligand **55b**, Eu^{3+} cation, and an aromatic β-diketonate anion [136]. Addition of F^-, acetate, HCO_3^-, and tartarate anions gave gradual luminescence changes at

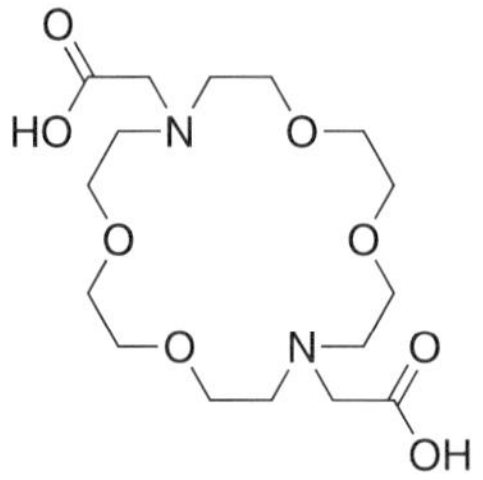

53

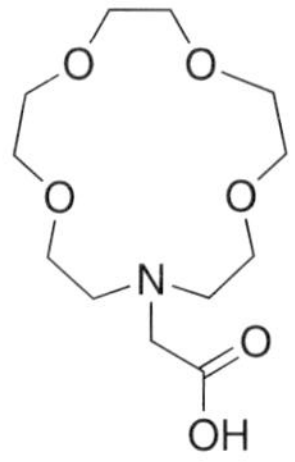

54

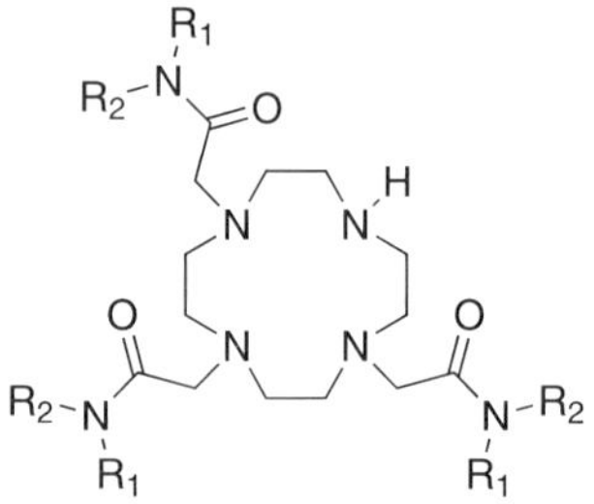

55a: R_1 = H ; R_2 = Me
55b: R_1 = R_2 = Me

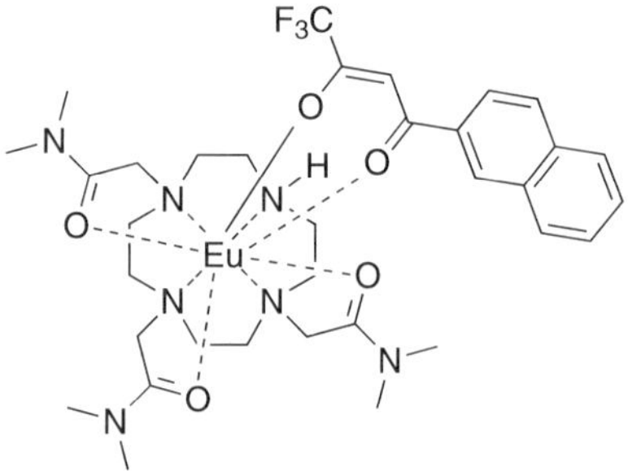

[Eu^{3+}–**55b**(β-diketonato)]$^{2+}$ complex

Fig. 20 Heptadentate ligands **55** and ternary Eu^{3+} complex with a β-diketonate ligand

neutral pH, while Br^-, PF_6^-, NO_2^-, and ClO_4^- anions induced no significant response. Parker et al. developed the sophisticated cyclen derivatives **56** and **57** bearing a phenanthridinium pendant arm as a sensitizer. The resulting Eu^{3+} complexes exhibited dual emission signals upon excitation of the antenna (at 320 nm) [130, 137, 138]. When Cl^-, Br^-, or I^- anion was added to the Eu^{3+} complex solution, both phenanthridinium fluorescence (at 405 nm) and europium luminescence (at 616 nm) decreased. A similar luminescence quenching was observed with Cl^- anion at pH 1.5–9, while the addition of bicarbonate, citrate, lactate, or phosphate anion had no such effect. Thus, this type of Eu^{3+} complexes is a potential candidate as Cl^--sensing material effective in water.

Yu and Parker prepared cyclen derivatives **58** and **59** bearing a pyridothioxanthone for the photosensitization of luminescent Eu^{3+} cation, which provided a 6- (N_4O_2) and a 7-coordination (N_5O_2) mode [139]. Addition of malate anion to a solution of Eu^{3+}–**58** increased the intensity of the $\Delta J = 2$ bands, but decreased that of $\Delta J = 0$ transition. Eu^{3+}–**59** exhibited a pronounced sensitivity for citrate anion [140]. Therefore, these were applicable in the ratiometric analysis or imaging of the aforementioned anions.

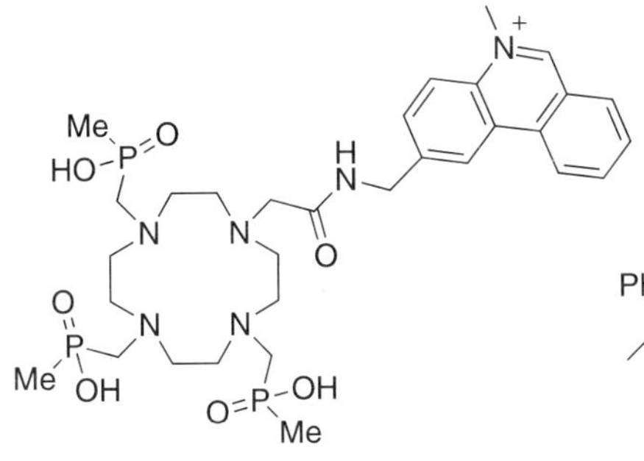

56 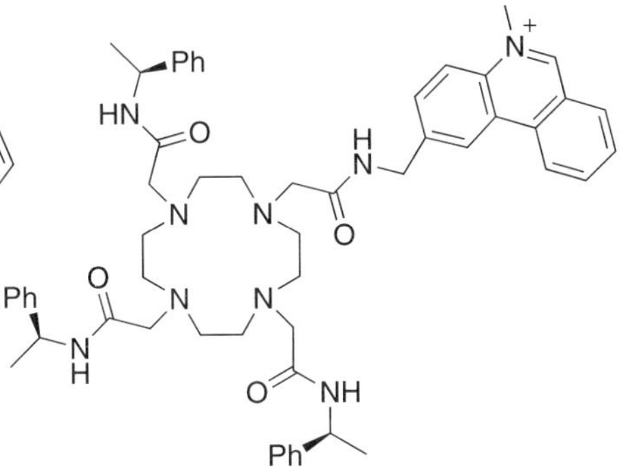 **57**

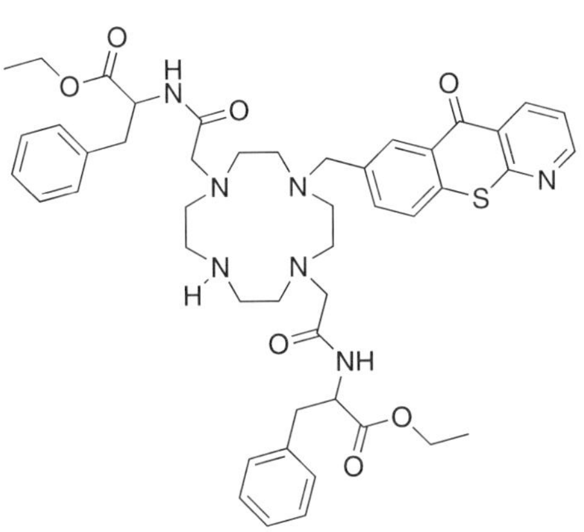

58 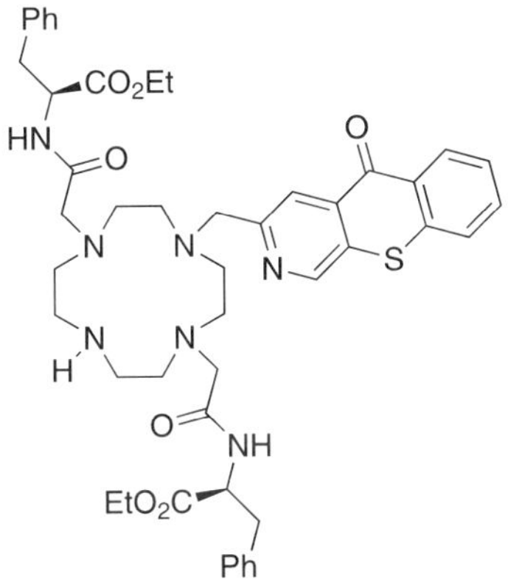 **59**

4.4
Lanthanide Complexes for NMR Sensing

Although MRI is practically applied in clinical fields, interesting examples of NMR sensing with lanthanide complexes were reported. Terreno et al. employed cyclen **60**–lanthanide complexes. This class of complexes yielded two stereogenic centers (Fig. 21): one generated from the two possible conformations of the four ethylenediamine groups of the cyclen ring ($\delta\delta\delta\delta$ or $\lambda\lambda\lambda\lambda$), and the other arising from the two possible orientations of the coordinating three pendant arms (Δ or Λ) [141]. **60**-Yb^{3+} complex exhibited sharp ^{1}H-NMR signals ($278 < T < 298$ K) indicating that two enantiomers dynamically exchanged in solution. Proton relaxation enhancement experiments revealed that Gd^{3+} and Yb^{3+} complexes with **60** had affinity following the order tar-

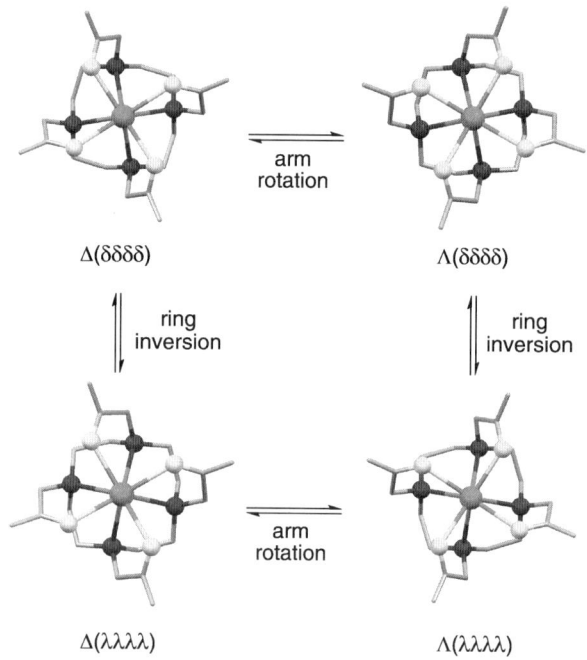

Fig. 21 Stereochemistry of cyclen–lanthanide complex with four pendant arms. Reprinted with permission from [119]. © (2002), American Chemical Society

tarate > lactate > malate > trifluorolactate (Fig. 22). The higher affinity was recorded with tartrate anion having two α-hydroxy-carboxylates. The ternary adducts between Yb^{3+}–**60** complex and these (S)-substrates exhibited sharper resonances with a split into two peaks of each signal. The intensity ratio of proton signals of the two diastereoisomers ($\Delta(\lambda\lambda\lambda\lambda)$ + (S)-substrate vs $\Lambda(\delta\delta\delta\delta)$ + (S)-substrate) provided a clear indication of the enantio-selective interaction:

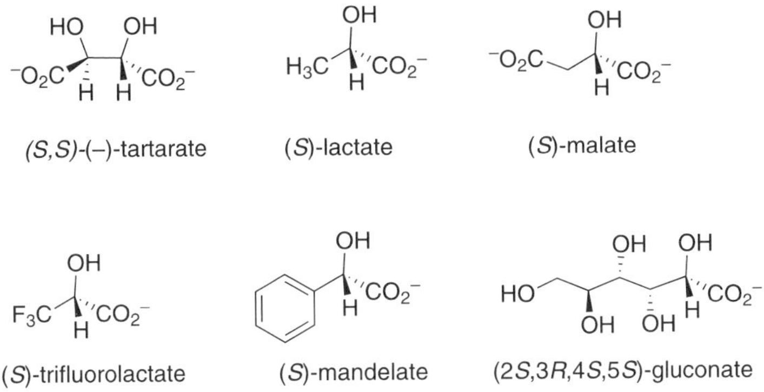

Fig. 22 Structures of (*S*)-isomers of α-hydroxy-carboxylate anions

1 : 1 for lactate and mandelate anions, 1.7 : 1 for malate anion, 2 : 1 for gluconate anion, 2.7 : 1 for tartrate anion, and 3 : 1 for trifluorolactate anion (Fig. 22). While lactate and mandelate anions coordinated the two enantiomers of Yb^{3+}–**60**, the other anions interacted preferentially with a specific enantiomer. The – CF_3, – OH, and – COOH moieties of the anions were thought to stabilize one diastereoisomer through the hydrogen bond formation.

Dickins et al. prepared Ho^{3+} and Yb^{3+} complexes with chiral cyclens **50a** and **61**, which had two vacant sites for external anions [142, 143]. Upon addition of an equivalent of (*S*)-lactate anion to a solution of Yb^{3+}–**50a** complex, two additional singlet ^{1}H-NMR resonances for CH and CH_3 protons of the guest appeared. Although CH_3 signals were broadened due to paramagnetic ion effect, the observed paramagnetic shifts suggested that the CH group of the lactate anion lay closer to the principal axis of the complex than its CH_3 group. This was interestingly consistent with the X-ray crystal data of the $\Lambda(\delta\delta\delta\delta)$ form. Addition of racemic lactate anions to a solution of Yb^{3+}–**50a** gave a 1 : 1 mixture of diastereomeric complexes. The CH and CH_3 resonances of the lactate anion were clearly resolved for the (*R*)-and (*S*)-diastereomers. A closely similar response was observed with related α-hydroxy acids, but mandelate anion offered no enantio-selectivity in complex formation. Yb^{3+}–**61** complex also provided a bidentate chelation mode for the lactate anion. The resonance of the lactate anion was clearly resolved for the (*R*)-and (*S*)-diastereoisomers, while the cyclen protons were also resolved. In the ^{1}H NMR spectra, the order of magnitude shifts of the axial ring protons was in the arrangement CO_3^{2-} < oxalate < citrate < acetate < PO_4^{3-}. These Yb^{3+} complexes further allowed additional CD and emission signals of the aforementioned anions.

Several heterocycle–lanthanide complexes provided attractive, imaginative, and suggestive approaches to anion sensing, in which ligand architecture based on heterocyclic chemistry played important roles. Cyclen-lanthanide complexes have several advantages of facile design and systematic synthesis, while the designed acyclic ligands also have promising applications for in vivo anion detection. The next practical goal should be design of heterocycle-lanthanide complexes capable of absorbing and emitting in the near-IR region, in which many heterocycles can work as useful building blocks.

5
Conclusion

This chapter summarized recent advances in the chemistry of the heterocycle-based receptors used in molecular recognition and supramolecular assembly systems. Because of their unique chemical and physical properties, the designed heterocyclic compounds provided specific recognition of targeted cationic or anionic guests. Since they varied widely in their structural, electronic, and interaction properties, their molecular recognition profiles were optimized through molecular architecture. In addition to the heterocyclic compounds, their lanthanide complexes also acted as effective receptors for specific anions. The synthetic strategies for these specific receptors further provided interesting extensions of heterocyclic chemistry to heterocyclic supramolecular chemistry. Since heterocyclic chemistry is one of the most established chemical sciences, its advances can offer more sophisticated molecular recognition and supramolecular assembly systems of the next generation.

Acknowledgements The authors express their thanks to the Ministry of Education, Culture, Sports, Science, and Technology of Japan for supporting their researches cited here (under Grant-in-Aid of Science Research, No. 16080217). Permissions to use Figs. 3, 4, 5, 18, and 21 and Table 1 from the American Chemical Society, Fig. 6 from the Royal Society of Chemistry, and Fig. 17 from the Rare Earth Society of Japan are also very much appreciated.

References

1. Atwood AL, Davies JED, MacNicol DD, Vögtle F (eds) (1996) Comprehensive supramolecular chemistry, vol 1, 2, and 6. Pergamon, New York
2. Sauvage JP (1990) Acc Chem Res 23:319
3. Fujita M, Ogura K (1996) Coord Chem Rev 148:249
4. Dutzler R, Campbell EB, MacKinnon R (2003) Science 300:108
5. Dutzler R, Campbell EB, Cadene M, Chait BT, MacKinnon R (2002) Nature 415:287
6. Clare JP, Ayling AJ, Joos J-B, Sisson AL, Magro G, Pérez-Payán MN, Lambert TN, Shukla R, Smith BD, Davis AP (2005) J Am Chem Soc 127:10739
7. Davis JT (2004) Angew Chem Int Ed 43:668
8. Boyd PDW, Reed CA (2005) Acc Chem Res 38:235
9. Sato H, Tashiro K, Shinmori H, Osuka A, Murata Y, Komatsu K, Aida T (2005) J Am Chem Soc 127:13086

10. Prince RB, Barnes SA, Moore JS (2000) J Am Chem Soc 122:2758
11. Garric J, Léger J-M, Huc I (2005) Angew Chem Int Ed 44:1954
12. Waki M, Abe H, Inouye M (2007) Angew Chem Int Ed 46:3059
13. Kumar S, Paul D, Singh H (2005) Adv Heterocycl Chem 89:65
14. Morohashi N, Narumi F, Iki N, Hattori T, Miyano S (2006) Chem Rev 106:5291
15. Ishibashi K, Tsue H, Tokita S, Matsui K, Takahashi H, Tamura R (2006) Org Lett 8:5991
16. Tsukube H, Yamada T, Shinoda S (2000) Ind Eng Chem Res 39:3412
17. Tsukube H, Shinoda S, Uenishi J, Hiraoka, Imakoga T, Yonemitsu O (1998) J Org Chem 63:3884
18. Tsukube H, Yamada T, Shinoda S (2004) J Alloys Compd 374:40
19. Tsukube H (1990) In: Inoue Y, Gokel GW (eds) Cation binding by macrocycles. Marcel Dekker, New York, p 487
20. Erickson SD, Still WC (1990) Tetrahedron Lett 31:4253
21. Dai Z, Xu X, Canary JW (2002) Chem Commun, p 1414
22. Kataoka Y, Paul D, Miyake H, Shinoda S, Tsukube H (2007) Dalton Trans 2784
23. Yamada Y, Shinoda S, Sugimoto H, Uenishi J, Tsukube H (2003) Inorg Chem 42:7932
24. Tsukube H (1996) Coord Chem Rev 148:1
25. Tsukube H, Yamada T, Shinoda S (2001) J Hetrocycl Chem 38:1401
26. van Unen D-J, Engbersen KFJ, Reinhoudt DN (2001) J Mol Catal B: Enzym 11:877
27. Itoh T, Takagi Y, Tsukube H (1997) J Mol Catal B: Enzym 3:259
28. Odell B, Earlam G (1985) J Chem Soc Chem Commun, p 359
29. Julian RR, Beauchamp JL (2001) Int J Mass Spectrom 613:210–211
30. Paul D, Suzumura A, Sugimoto H, Teraoka J, Shinoda S, Tsukube H (2003) J Am Chem Soc 125:11478
31. Hamuro Y, Calama MC, Park HS, Hamilton AD (1997) Angew Chem Int Ed Engl 36:2680
32. Oshima T, Goto M, Furusaki S (2002) Biomacromolecules 3:438
33. Paul D, Miyake H, Shinoda S, Tsukube H (2006) Chem Eur J 12:1328
34. Shinoda S, Ohashi M, Tsukube H (2007) Chem Eur J 13:81
35. Tsukube H, Suzuki Y, Paul D, Kataoka Y, Shinoda S (2007) Chem Commun, p 2533
36. Park CH, Simmons HE (1968) J Am Chem Soc 90:2431
37. Schmidtchen FP, Müller GJ (1980) Chem Ber 113:864
38. Schmidtchen FP, Müller GJ (1984) Chem Commun, p 1115
39. Dietrich B, Guilhem J, Lehn J-M, Pascard C, Sonveaux E (1984) Helv Chim Acta 67:91
40. Hosseini H, Lehn J-M (1986) Helv Chim Acta 69:587
41. Bianch A, Bowman-James K, García-España (eds) (1997) Supramolecular chemistry of anions. Wiley-VCH, New York
42. Beer PD, Gale PA (2001) Angew Chem Int Ed 40:486
43. Sessler JL, Gale PA, Cho W-S (2006) Anion Receptor Chemistry. RSC Publishing, Cambridge
44. Bryantsev VS, Hay BP (2005) J Am Chem Soc 127:8282
45. Berryman OB, Bryantsev VS, Stay DP, Johnson DW, Hay BP (2006) J Am Chem Soc 129:48
46. Choi K, Hamilton AD (2003) Coord Chem Rev 240:101
47. García-España E, Díaz P, Llinares JM, Bianchi A (2006) Coord Chem Rev 250:2952
48. Wichmann K, Antonioli B, Söhnel T, Wenzel M, Gloe K, Price JR, Lindoy LF, Blake AJ, Schröder M (2006) Coord Chem Rev 250:2987
49. Tabushi I, Kumura Y, Yamamura K (1978) J Am Chem Soc 100:1304
50. Tabushi I, Kumura Y, Yamamura K (1981) J Am Chem Soc 103:6486

51. Cramer RE, Fermin V, Kuwabara E, Kirkup R, Selman M, Aoki K, Adeyemo A, Yamazaki H (1991) J Am Chem Soc 113:7033
52. Cramer RE, Carrié MJ (1990) Inorg Chem 29:3902
53. Cramer RE, Mitchell KA, Hirazumi AY, Smith SL (1994) J Chem Soc Dalton Trans 563
54. Cramer RE, Smith DW, VanDoorne W (1998) Inorg Chem 37:5895
55. Shinoda S, Tadokoro M, Tsukube H, Arakawa R (1998) Chem Commun, p 181
56. Cabildo P, Sanz D, Claramunt RM, Rourne SA, Alkorta I, Elguero J (1999) Tetrahedron 55:2327
57. Alcalde E, Alvarez-Rúa C, García-Granda S, García-Rodriguez E, Mesquida N, Pérez-García L (1999) Chem Commun, p 295
58. Ramos S, Alcalde E, Doddi G, Mencarelli P, Pérez-García L (2002) J Org Chem 67:8463
59. Chellappan K, Singh NJ, Hwang I-C, Lee JW, Kim KS (2005) Angew Chem Int Ed 44:2899
60. Sato K, Arai S, Yamagishi T (1999) Tetrahedron Lett 40:5219
61. Ihm H, Yun S, Kim HG, Kim JK, Kim KS (2002) Org Lett 4:2897
62. Yoon J, Kim SJ, Singh NJ, Kim KS (2006) Chem Soc Rev 35:355
63. Abouderbala LO, Belcher WJ, Boutelle MG, Cragg PJ, Dhaliwal J, Fabre M, Steed JW, Turner DR, Wallace KJ (2002) Chem Commun, p 358
64. Wallace KJ, Belcher WJ, Turner DR, Syed KF, Steed JW (2003) J Am Chem Soc 125:9699
65. Ilioudis CA, Tocher DA, Steed JW (2004) J Am Chem Soc 126:12395
66. Gale PA (2000) Coord Chem Rev 199:181
67. Gale PA (2001) Coord Chem Rev 213:79
68. Gale PA (2003) Coord Chem Rev 240:191
69. Gale PA, Quesada R (2006) Coord Chem Rev 250:3219
70. Kang SO, Hossain MA, Bowman-James K (2006) Coord Chem Rev 250:3038
71. Kang SO, Begun RA, Bowman-James K (2006) Angew Chem Int Ed 45:7882
72. Sessler JL, Cyr MJ, Lynch V, McGhee E, Ibers JA (1990) J Am Chem Soc 112:2810
73. Shionoya M, Furuta H, Lynch V, Harriman A, Sessler JL (1992) J Am Chem Soc 114:5714
74. Gale PA, Sessler JL, Král V, Lynch V (1996) J Am Chem Soc 118:5140
75. Sessler JL, An D, Cho W-S, Lynch V (2003) J Am Chem Soc 125:13646
76. Sessler JL, An D, Cho W-S, Lynch V (2003) Angew Chem Int Ed 42:2278
77. Sessler JL, An D, Cho W-S, Lynch V, Marquez M (2005) Chem Eur J 11:2001
78. Gale PA, Camiolo S, Tizzard GJ, Chapman CP, Light ME, Coles SJ, Hursthouse MB (2001) J Org Chem 66:7849
79. Gale PA (2006) Acc Chem Res 39:465
80. Chmielewski MJ, Charon M, Jurczak J (2004) Org Lett 6:3501
81. Maeda H, Ito Y (2006) Inorg Chem 45:8205
82. Maeda H, Kusunose Y (2005) Chem Eur J 11:5661
83. Chang K-L, Moon D, Lah MS, Jeong K-S (2005) Angew Chem Int Ed 44:7926
84. Katayev EA, Boev NV, Khrustalev VN, Ustynyuk YA, Tananaev IG, Sessler JL (2007) J Org Chem 72:2886
85. Martínez-Máñez R, Sancenón F (2003) Chem Rev 103:4419
86. Suksai C, Tuntulani T (2003) Chem Soc Rev 32:192
87. Black CB, Andrioletti B, Try AC, Ruiperez C, Sessler JL (1999) J Am Chem Soc 121:10438
88. Nishiyabu R, Anzenbacher P jr (2005) J Am Chem Soc 127:8270

89. Evans LS, Gale PA, Light ME, Quesada R (2006) Chem Commun, p 965
90. Nguyen BT, Anslyn EV (2006) Coord Chem Rev 250:3118
91. Metzger A, Anslyn EV (1998) Angew Chem Int Ed 37:649
92. Piatek AM, Bomble YJ, Wiskur SL, Anslyn EV (2004) J Am Chem Soc 126:6072
93. Wiskur SL, Anslyn EV (2001) J Am Chem Soc 123:10109
94. Gale PA, Twyman LJ, Handlin CI, Sessler JL (1999) Chem Commun, p 1851
95. Atilgan S, Akkaya EU (2004) Tetrahedron Lett 45:9269
96. Beer PD, Hayes EJ (2003) Coord Chem Rev 240:167
97. Sigel A, Sigel H (eds) (2003) Metal ions in biological systems, vol 40. Marcel Dekker, New York
98. Parker D (2004) Chem Soc Rev 33:156
99. Tsukube H, Shinoda S (2002) Chem Rev 102:2389
100. Shinoda S, Miyake H, Tsukube H (2005) Handbook on the physics and chemistry of rare earths 35:273
101. Di Bari L, Salvadori P (2005) Coord Chem Rev 249:2854
102. Riehl JP, Muller G (2005) Handbook on the physics and chemistry of rare earths 34:289
103. Wenzel TJ, Wilcox JD (2003) Chirality 15:256
104. Rothchild R, Wyss H (1994) Spectrosc Lett 27:225
105. Inamoto A, Ogasawara K, Omata K, Kabuto K, Sasaki Y (2000) Org Lett 2:3543
106. Di Bari L, Lelli M, Pintacuda G, Salvadori P (2002) Chirality 14:265
107. Aime S, Botta M, Parker D, Williams JAG (1995) J Chem Soc Dalton Trans 2259
108. Aime S, Botta M, Crich SG, Terreno E, Anelli PL, Uggeri F (1999) Chem Eur J 5:1261
109. Corsi DM, van Bekkum H, Peters JA (2000) Inorg Chem 39:4802
110. Lakowicz JR (1999) Principles of fluorescence spectroscopy, 2nd edn. Springer, Berlin Heidelberg New York
111. Valeur B (2002) Molecular fluorescence: principles and applications. Wiley-VCH, Weinheim
112. Douce L, Charbonnière L, Cesario M, Ziessel R (2001) New J Chem 25:1024
113. Cho EJ, Moon JW, Ko SW, Lee JY, Kim SK, Yoon J, Nam KC (2003) J Am Chem Soc 125:12376
114. Richardson FS (1982) Chem Rev 82:541
115. Comby S, Bünzli J-CG (2007) Handbook on the physics and chemistry of rare earths 37:217
116. Reuben J (1979) J Chem Soc Chem Commun, p 68
117. Reuben J (1980) J Am Chem Soc 102:2232
118. Bünzli J-CG (1989) In: Bünzli J-CG, Choppin GR (eds) Lanthanide probes in life, chemical and earth sciences. Elsevier, Amsterdam, p 219
119. Parker D, Dickins RS, Puschmann H, Crossland C, Howard JAK (2002) Chem Rev 102:1977
120. Mathis G (1995) Clin Chem 41:1391
121. Dickson EFG, Pollack A, Diamandis EP (1995) J Photochem Photobiol B 27:3
122. Elbanowski M, Makowska B (1996) J Photochem Photobiol A 99:85
123. Faulkner S, Mattews JL (2003) In: Ward MD (ed) Comprehensive coordination chemistry II, vol 9. Elsevier, Amsterdam, p 913
124. Balzani V, Sabbatini N, Scandola F (1986) Chem Rev 86:319
125. Montalti M, Prodi L, Zaccheroni N, Charbonnière L, Douce L, Ziessel R (2001) J Am Chem Soc 123:12694
126. Charbonnière LJ, Ziessel R, Montalti M, Prodi L, Zaccheroni N, Boehme C, Wipff G (2002) J Am Chem Soc 124:7779

127. Yamada T, Shinoda S, Tsukube H (2002) Chem Commun, p 1218
128. Best MD, Anslyn EV (2003) Chem Eur J 9:51
129. Dickins RS, Gunnlaugsson T, Parker D, Peacock RD (1998) Chem Commun, p 1643
130. Bruce JI, Dickins RS, Govenlock LJ, Gunnlaugsson T, Lopinski S, Lowe MP, Parker D, Peacock RD, Perry JJB, Aime S, Botta M (2000) J Am Chem Soc 122:9674
131. Mameri S, Charbonnière LJ, Ziessel RF (2003) Synthesis 17:2713
132. Charbonnière LJ, Schurhammer R, Mameri S, Wipff G, Ziessel RF (2005) Inorg Chem 44:7151
133. Ziessel RF, Charbonnière LJ, Mameri S, Camerel F (2005) J Org Chem 70:9835
134. Magennis SW, Craig J, Gardner A, Fucassi F, Cragg PJ, Robertson N, Parsons S, Pikramenou Z (2003) Polyhedron 22:745
135. Gunnlaugsson T, Harte AJ, Leonard JP, Nieuwenhuyzen M (2002) Chem Commun, p 2134
136. Leonard JP, dos Santos CMG, Plush SE, McCabe T, Gunnlaugsson T (2007) Chem Commun, p 129
137. Parker D, Senanayake K, Williams JAG (1997) Chem Commun, p 1777
138. Parker D, Senanayake K, Williams JAG (1998) J Chem Soc Perkin Trans 2:2129
139. Yu J, Parker D (2005) Eur J Org Chem 4249
140. Parker D, Yu J (2005) Chem Commun, p 3141
141. Terreno E, Botta M, Fedeli F, Mondino B, Milone L, Aime S (2003) Inorg Chem 42:4891
142. Dickins RS, Love CS, Puschmann H (2001) Chem Commun, p 2308
143. Dickins RS, Badari A (2006) Dalton Trans 3088
144. Yamada T, Shinoda S, Kikawa K, Ichimura A, Teraoka J, Takui T, Tsukube T (2000) Inorg Chem 39:3049

Syntheses and Properties of Crownophanes

S. Inokuma · M. Ito · J. Nishimura (✉)

Department of Chemistry and Chemical Biology, Graduate School of Engineering, Gunma University, Tenjin-cho, Kiryu 376-8515 Gunma, Japan
nisimura@chem-bio.gunma-u.ac.jp

1	Introduction	44
2	Benzene Ring(s) Containing Crownophanes	46
3	Naphthalene Ring(s) Containing Crownophanes	53
4	Other Condensed Polyaromatic Ring(s) Containing Crownophanes	55
4.1	Fluorenone and Stilbene Ring(s) Containing Crownophanes	55
4.2	Anthracene Ring(s) Containing Crownophanes	56
4.3	Pyrene Rings Containing Crownophanes	57
5	Heteroaromatic Ring Containing Crownophanes	58
5.1	Pyridine Ring(s) Containing Crownophanes	58
5.2	Bipyridine Ring(s) Containing Crownophanes	59
5.3	Phenanethroline Ring(s) Containing Crownophanes	60
6	Catenanes and Rotaxanes	62
7	Concluding Remarks	67
References		68

Abstract A variety of crownophanes possessing both crown ether and cyclophane moieties are reviewed and their specific complexation are described. Aromatic nuclei including benzene, naphthalene, anthracene, pyridine, and other condensed polyaromatic rings are dealt with as components of crownophanes. Crownophanes containing nitrogen and sulfur atoms as ligating parts in the tethers of the aromatics are also described. Oxygen-containing crownophanes show high affinity toward alkali and alkaline metal cations with high selectivity. For example, three-bridged crownophane **25** extracted Li^+ with excellent selectivity and high efficiency in the competitive system containing Na^+ and K^+. Four-bridged crownophane **26b** showed extraordinarily high selectivity toward Na^+ with high efficiency compared with commercially available 15-crown-5 and benzo-15-crown-5. Crownophane **1** having pyridine nuclei as secondary ligating sites on the benzene rings efficiently complexed Ag^+ ion with perfect selectivity. Di(*p*-phenylene) crown **60**, 1,5-dinaphtho crown **61**, and 1,5-naphtho-*p*-phenylene crown **62** have been widely employed for donor components of catenanes and rotaxanes. Characteristic behaviors of typical and important crownophanes including above-mentioned ones are summarized in Table 1.

Keywords Catenane · Crown ether · Crownophane · Cyclophane · Rotaxane

1
Introduction

Since the synthesis and properties of crown ethers like dibenzo-18-crown-6 were published by Pedersen in 1967 [1] and developed by Cram, Lehn and other eminent researchers in the 1970s, the parent hosts have been extensively applied to the supramolecular chemistry in a variety of ways [2–4].

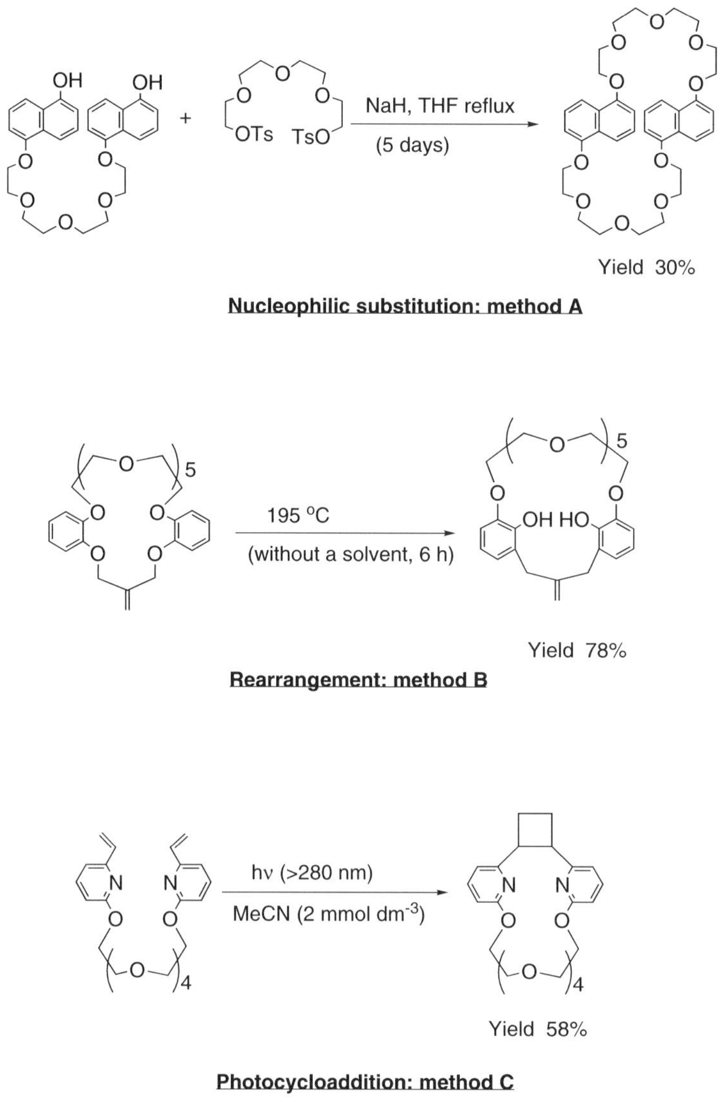

Scheme 1 Cyclization methodologies for crownophanes

Syntheses and Properties of Crownophanes 45

In this highly potential materials group there is a characteristic (but rather small) subgroup named crownophanes. The name was coined from the crown ether and the cyclophane [5, 6], which is another major group covering many materials listed in the supramolecular chemistry. The crownophanes literally means "crown ether moieties having cyclophane moieties" [7–13]. Although the meaning of the former is understandable, the latter needs an explana-

Table 1 Typical and important crownophanes and their properties

Category of crownophanes	Remarks
Ag^+-binding crownophane	Crownophane **1** efficiently complexed Ag^+ ion with perfect selectivity
Zn^{2+}-complexing crownophanes	Crownophanes **2a** and **2b** efficiently complexed Zn^{2+} ion
Li^+-binding crownophane	Crownophane **25** showed extraordinarily high extraction selectivity for Li^+ toward Na^+ and K^+
Na^+-binding crownophane	Crownophane **26b** exhibited selective Na^+-extraction compared with commercially available crown ethers
K^+-binding crownophane	Crownophane **27b** extracted K^+ extensively with high selectivity compared with commercially available 18-crown-6
Receptors for neutral molecules	Crownophane **31b** formed a 1 : 1 complex with urea at $-50\,°C$
	Crownophane **33** complexed H_2CO_3 generated from H_2O and CO_2
Anion-binders	Crownophanes **34a** and **34b** recognized $H_2PO_4^-$
	Crownophane **67** Complexed 1,3,5-benzene-tricarboxylate anion
Ammonium-binding crownophanes	Crownophanes **38a** and **38b** strongly complexed dibenzylammonium cation than the corresponding dibenzocrown ethers
Water-soluble crownophane	Crownophane **42** were cationic water-soluble pyrenophane with strong recognition toward anionic arenas including nucleotides in water
Rotaxanes and catenanes	Crownophanes **60–62** were of donors components of rotaxanes and catenanes
	Rotaxanes **72** were prepared via covalent bond formation
Alkali metal-binding rotaxanes	Rotaxane **74** had Li^+-binding ability
	Rotaxane **78** was the first [1]rotaxane having selective Li^+-complexing ability
	[3]Rotaxane **79** formed 1 : 2 Li^+-complex and 1 : 1 Cs^+-complex

tion. Cyclophane moieties are defined as those possessing some overlapping between aromatic ring faces or between an aromatic ring face and tether, as exemplified in [2.2]*para*cyclophane and in [5]*meta*cyclophane, respectively. Moreover, even though the overlapping area is quite narrow and/or instantaneous during conformational movement, such a compound can also be called a cyclophane.

Generally speaking, when the crown ethers are modified by adding the cyclophane moieties, they become naturally more hydrophobic, open narrower orifice by pinching with cyclophane part, and become more functional than the original crown ether because the cyclophane part provides a variety of platforms or sites for functional groups assisting their supramolecular properties such as lariats, etc.

For the preparation of crownophanes, two major strategies are usually taken: one is the method from linear precursors using [2+2] photocycloaddition or tandem Claisen rearrangement, and the other is of the addition of oligo-oxyethylene tethers to the preorganized aromatic motifs such as *para*cyclophane, 1,8-naphthalenes, and so on. Of course, almost all the conventional bond-formation methodologies are applied to the purpose, especially facile heteroatom-to-carbon bond formation reactions (method A in Scheme 1). Some template effects are also occasionally used in order to make cyclization practical. In Scheme 1, three useful syntheses are summarized.

In this chapter the scope and limitation of crownophanes are reviewed, focusing on supramolecular functions as ionophores and receptors. Most are designed for the recognition of metal cations and some organic cations, and done partly for neutral organic materials. Their structures are classified into five groups from benzene nuclei to heteroaromatic nuclei. In the individual section their supramolecular properties are summarized and the properties are mainly focused on metal ion-binding and chiral and/or achiral neutral molecule complexation. Rotaxanes and catenanes are also included in this chapter. Their preparation almost always accompanies some supramolecular complexation. Some typical and important crownophanes are summarized in Table 1.

2
Benzene Ring(s) Containing Crownophanes

Inokuma et al. have developed a synthetic method of crownophanes by intramolecular [2+2] photocycloaddition of linear precursors (method C). It was anticipated that they would bring about the specific complexing ability due to their crown ether-cyclophane hybrid structures. This photocycloaddition has been applied to the synthesis of crownophanes possessing hydroxyl groups on the aromatic nuclei. They are easily derived to such a lariat type of crownophanes as **1**. In the liquid–liquid extraction of heavy metal cations, crownophane **1** showed perfect selectively and high efficiency toward Ag$^+$ [14].

On the other hand, crownophanedicarboxylic acids **2a–c** showed low affinity toward alkali metal cations, though both **2b** and **2c** showed extraordinary high affinity toward Zn^{2+} ion among heavy metal cations examined [15].

A conformationally restricted polyoxygenated crownophane **3** was prepared by McMurry pinacol reaction, though their recognition has not yet been examined [16].

1,3,5-Triaroylbenzene-based crownophane **4** was prepared from regioselective cross-benzannulation between bis(arylethynyl) ketone and enaminone [17].

The crownophanes **5, 6, 9** and **10** [18] showed some affinity to alkali metal and ammonium cations in gas phase (ESI mass spectrometry), but failed to

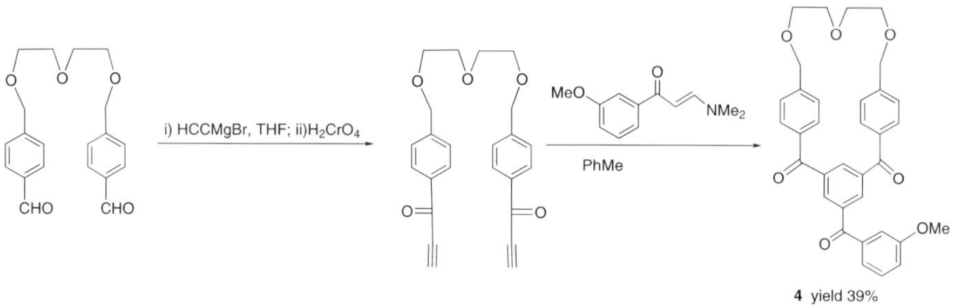

Scheme 2 Preparation of compound **4**

exhibit ionophoric properties in solution. For example, *para*crownophanes **5** and **6** interacted almost equally with Na$^+$, K$^+$, and Cs$^+$ in gas phase, while they hardly interacted to Li$^+$ or NH$_4$$^+$. In contrast, *meta*-isomer **10** exhibited similar affinity for all the cations. Crownophane **9** showed a preference for Na$^+$ over all other cations. Although there is no information regarding the site of cation–crownophane interaction, it seems that structural differences on the number of ether oxygen atoms, position isomerism regarding phenoxycarbonyl moieties are important factors that affect the binding abilities of **5–10** in the gas phase.

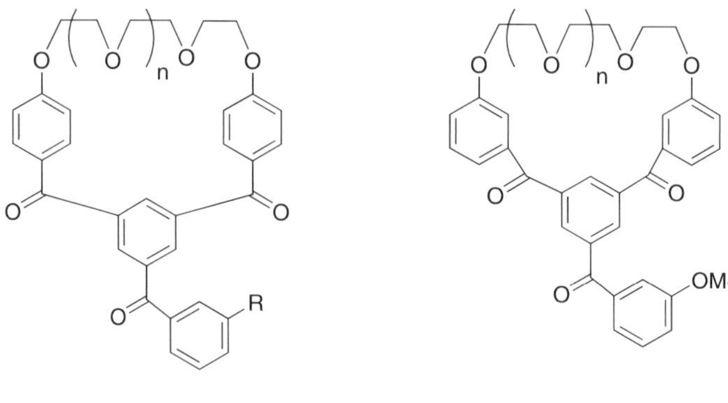

5 n = 1 R = OMe
6 n = 2 R = OMe
7 n = 2 R = H
8 n = 3 R = H

9 n = 1
10 n = 2

Hiratani et al. prepared a new type of crownophane **11** and **12** having two hydroxyl groups by using tandem Claisen rearrangement (method B) [19] and found that they form stable complexes with water (1 : 1 stoichiometry) [20]. The crownophane synthesis has become a unique and excellent method for the rotaxane synthesis described later Sect. 7.

11a-c
R = -(CH$_2$CH$_2$O)$_n$CH$_2$CH$_2$-
a n = 3
b n = 4
c n = 5

12 R = (OCH$_2$CH$_2$)$_3$- / (OCH$_2$CH$_2$)$_3$-

Crownophanes having sulfur atom(s) in the polyether linkage, thiacrownophanes **13–17**, were conveniently prepared from linear precursors (method C). These compounds showed high extractability toward Ag$^+$ [21, 22].

Inokuma et al. have prepared nitrogen-containing crownophanes, azacrownophane **18** and cryptocrownophane **19**, by the photocycloaddition (method C) in the presence of γ-cyclodextrin in aqueous phase, employing the repression of amino group-quenching effect by the inclusion of styrene moieties in γ-cyclodextrin. In the liquid–liquid extraction, azacrownophane

18 and cryptocrownophane 19 showed moderate affinity to Ag^+ and Pb^{2+} cations [23, 24]. Thus, the crownophane synthesis by the intramolecular [2+2] photocycloaddition of styrene derivatives is widely applied and recognized as one of the powerful methods for functionalization of heterocrownophanes.

Electrochemical properties of Wurster's crownophanes 20 and 21 were determined by cyclic voltammetry. The smaller crownophanes showed no electrochemical response to alkali metal cations, whereas 20c showed modest selectivity for alkaline earth metal cations and ammonium cation in the order $NH_4^+ < Ca^{2+} < Sr^{2+} < Ba^{2+}$ [25].

20a n = 1
 b n = 2
 c n = 3

21a n = 1
 b n = 2
 c n = 3

The complexation with Pt(II) for traditional crown 22 and Wurster's thiacrownophanes 23 was investigated by various techniques including ^{1}H NMR spectroscopy, electrospray mass spectrometry, cyclic voltammetry, and single crystal X-ray analysis. The crownophane geometry was proved to form unstable endocyclic complexes with Pt(II), compared with the traditional nest crown geometry [26].

22

23

Benzidine derivatives 24 strapped with a polyether unit at the 2,2′-positions was prepared by benzidine rearrangement of N,N′-diarylhydrazide derivatives in 45–47% yield. Characterizations were made on acetamide derivatives of 24, because compounds 24 were not purified enough by silica-gel column chromatography due to the interference of unidentified decomposed byproducts [27].

Rigid three-bridged crownophanes 25, e.g., crownpaddlanes possessing two cyclobutane rings were efficiently and selectively prepared (method C). Their complexing abilities toward Li^+, Na^+, and K^+ were evaluated by solid–liquid extractions. Crownpaddlanes 25 exclusively and quantitatively ex-

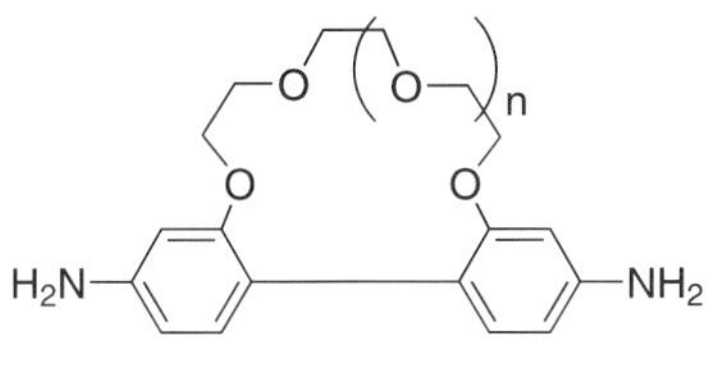

24 n = 2, 3, 4

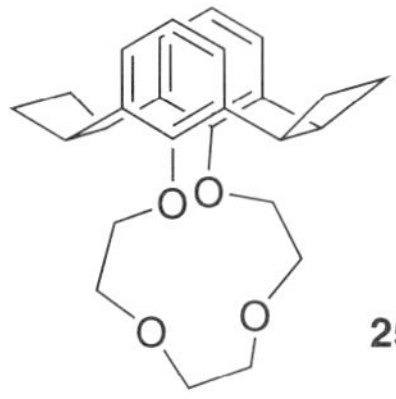

25

tracted Li$^+$ in single solid–liquid extraction. Upon a competitive extraction, **25** showed a higher selectivity toward Li$^+$ than toward Na$^+$ and K$^+$ (percent extraction for Li$^+$/percent extraction for Na$^+$ = 610, percent extraction for Li$^+$/percent extraction for K$^+$ = 980). The structural factors that influenced the complexation of the crownopaddlane were examined by X-ray crystallographic analysis [28]. The cavity diameter of **25** was found to be ca. 1.22 Å, which is quite appropriate for binding to Li$^+$. Furthermore, the cyclobutane blades of **25** may act as a steric barrier to 2 : 1 sandwich complexation.

Almost perfectly rigid four-bridged crownophane **26b** was prepared in 52% yield with the addition of NaBF$_4$ in the photoreaction system [29]. As this template effect suggests, **26b** showed extraordinarily high Na$^+$-selectivity with high efficiency in the liquid–liquid extraction of alkali metal picrates, while compound **26a** having four ethereal oxygen atoms did not extract any alkali metal cations in this system. The high Na$^+$-selectivity of **26b** was further clarified by the equilibrium stability constants (log K_a) for Na$^+$ (5.85) and K$^+$ (2.91) in acetonitrile solution. The log K_a value for Na$^+$ is 1000 times larger than that of commercially available benzo-15-crown-5. The complexation of **26b** to Na$^+$ cation was also examined by X-ray crystallography. It

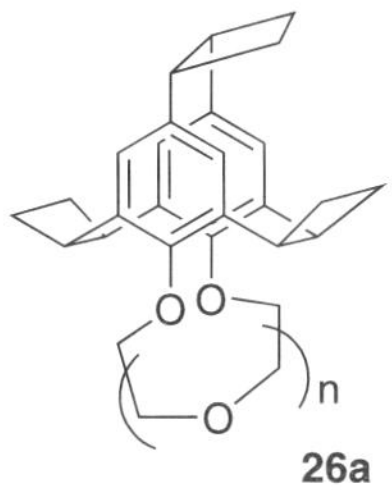

26a n = 2
b n = 3
c n = 4

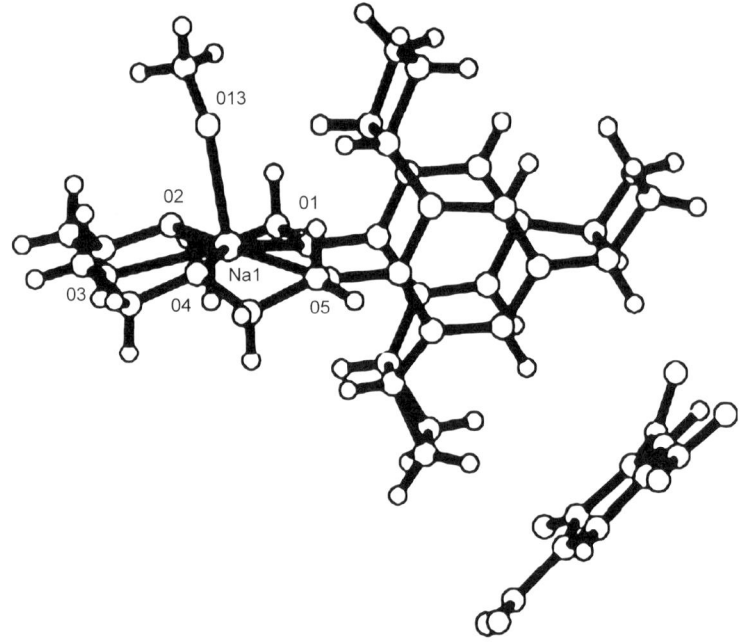

Fig. 1 Crystal structure of sodium picrate-**26b** complex [29]: selected bond distances (Å) and angles (deg): Na–O(1) 2.45; Na–O(2) 2.37; Na–O(3) 2.49; Na–O(4) 2.38; Na–O(5) 2.39; O(1)–Na–O(2) 70.4; O(2)–Na–O(3) 68.3, O(3)–Na–(4) 69.1; O(4)–Na–O(5) 71.7; O(5)–Na–O(1) 80.1 [29]

has perfectly layered aromatic nuclei that prevents itself from forming any sandwich-type complexes.

Crownopaddlane **26c** bearing six ethereal oxygen atoms also more efficiently and selectively extracted alkali metal cations, compared with 18-crown-6 derivatives [29]. The intramolecular photocycloaddition was also applied to prepare three- and four-bridged crownophanes. The yields were moderate or excellent (52%, 93%) in spite of one-pot reactions. The cyclobutane rings acted to not only make the phanes rigid but also control the complexation behavior due to their bulkiness.

Xu et al. have prepared a dithia[3.3]*meta*cyclophane **27a** bridged by a tetra(ethylene glycol) (method A) and found to form complexes with alkali metal cations in the affinity order of $Na^+ > K^+ > Rb^+ > Cs^+$. The $NaClO_4$-MeOH complex of the phane formed a hydrogen-bonded polymer in the solid phase, which is a supramolecular assembly stabilized via an intermolecular S···H–C (benzylic proton) interaction [30]. A dithia[3.3]*meta*cyclophane **27b** strongly complexed with K^+ cation. Phane **27b** formed a novel one-dimensional coordination polymer or supramolecular assembly with $KClO_4$ using anions as the linkers, while the complex of the phane with $NaClO_4$ took a dimeric structure [31].

Syntheses and Properties of Crownophanes 53

27a n = 1
b n = 2

The complexing behavior of dithia[n.3.3](1,3,5)crownophanes **28a–c** exhibits an unusual ion-selectivity due to so-called "breathing" process of the dithia[3.3]*meta*cyclophane moiety. The process is of induced fit. Thus, when a small ion comes, it shrinks the cavity, but when a large ion comes, it enlarges the cavity. The crown cavity of phane **28a** is thought to be too large for Li^+ but too small for the other alkali metals. The breathing flexibility in phane **28a** may be restricted due to the relatively short polyether linkage. For phane **28b**, the order of the association constants is $K^+ > Na^+ > Rb^+ > Cs^+$. For phane **28c**, the order of the association constants is $Cs^+ > Rb^+ > K^+ > Na^+$. This breathing mechanism is also supported by X-ray crystallographic analysis [32].

28a n = 0
b n = 1
c n = 2

3
Naphthalene Ring(s) Containing Crownophanes

Without high-dilution techniques, compounds **29** and **30** were obtained in overall ca. 30% yield (method A). Crownophanes **29** and **30** with 28–150 atoms were isolated. It was found by X-ray structural analysis that compound **29d** ($n = 1$) formed a complex with CH_2Cl_2 [33].

Hiratani et al. synthesized chiral crownophanes **31a** and **31b** having a binaphthyl unit and two naphthol units via tandem Claisen rearrangement (method B) in quantitative yields. From the association constants of the crownophanes with the enantiomers of phenylaranine, phenylglycinol, and phenylaraninol determined by ^1H NMR titration, crownophane **31a** has a chiral recognizability for the (R)-form of phenylglycinol over the (S)-form [34].

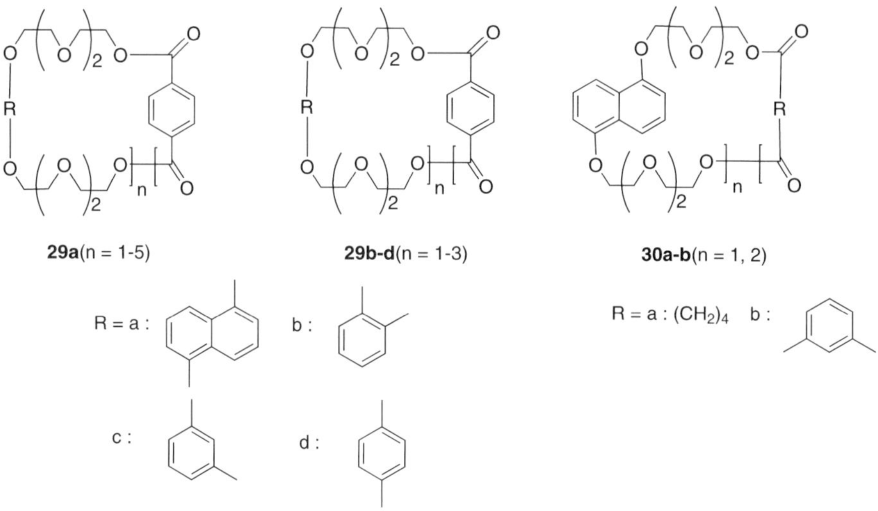

29a(n = 1-5)   **29b-d**(n = 1-3)   **30a-b**(n = 1, 2)

R = a: b: c: d:

R = a : (CH$_2$)$_4$ b :

It was found by ^{1}H NMR analysis that phane **31b** formed 1 : 1 complex with the urea at −50 °C [35].

They also synthesized crownophanes **32** having two hydroxyl groups in addition of two naphthalene rings in high yields via Claisen rearrangement [36].

Crownophane **33** having two amido groups in a polyether linkage in addition of two naphthol moieties was prepared by the same method. Phane **33** gives a stable 1 : 1 complex with carbonic acid formed from carbon dioxide and water at room temperature [37–39].

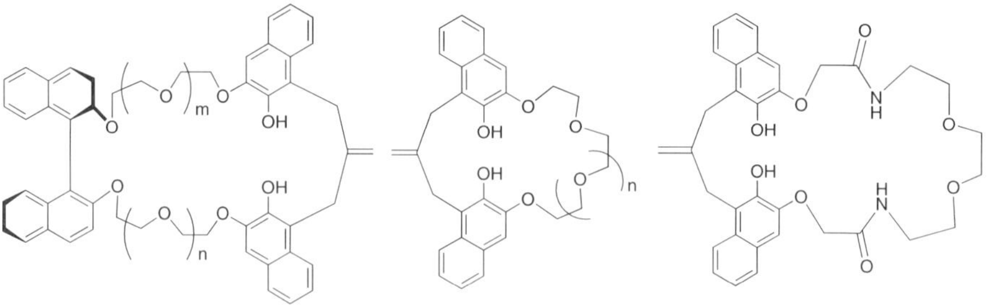

31a m = 1, n = 1
  b m = 2, n = 2

32a n = 1
  b n = 2
  c n = 3

33

Amindocrownophanes **34a** and **34b**, composed of 28-membered ring having two hydroxy groups, two amide groups, and naphthalene rings were prepared by method B from compounds **35a** and **35b**, respectively. It is strange that compounds **34** recognized anions in the following order; $H_2PO_4^- > F^- > CH_3COO^- > Cl^- \gg Br^-$ and I^-, whereas not only compounds **35** and **36** hav-

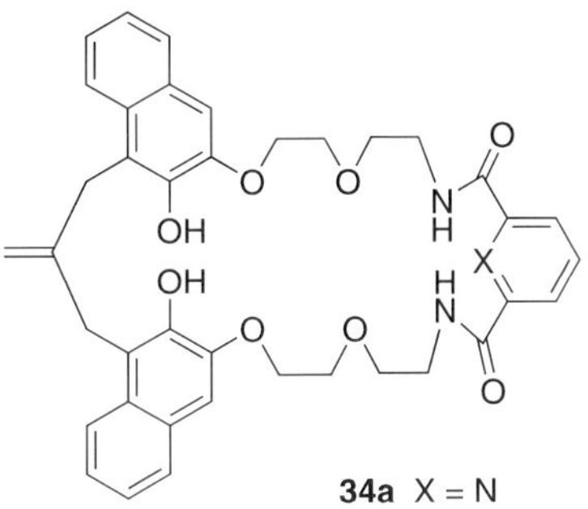

34a X = N
b X = CH

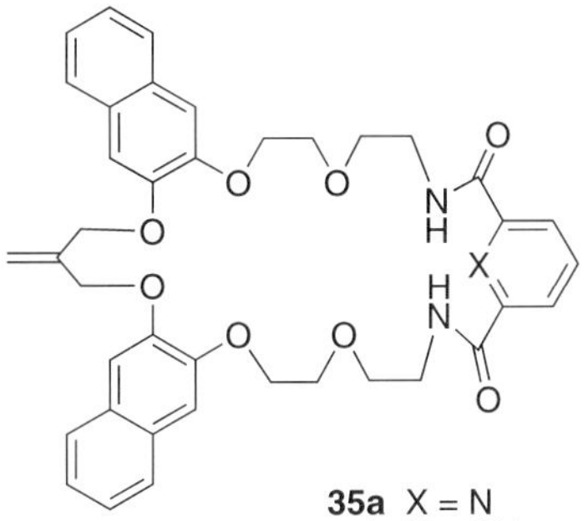

35a X = N
b X = CH

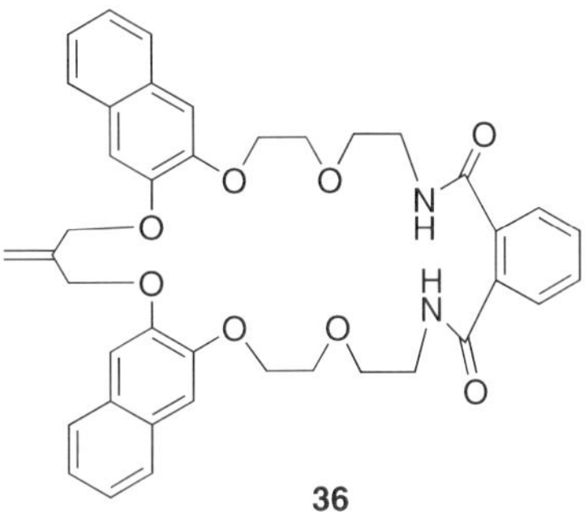

36

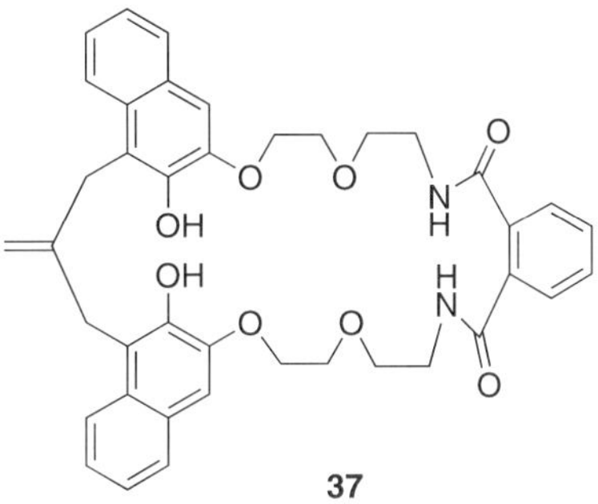

37

ing no hydroxy group but also compound **37** having 27-membered ring have no ability for anion recognition. Hence, amide groups, hydroxy groups, and *m*-phenylene or 1,6-pyridyl rigid moiety synergically play an important role for recognition ability [40].

Thus, the tandem Claisen rearrangement developed by Hiratani et al. is an elegant method to prepare functional crownophanes.

4
Other Condensed Polyaromatic Ring(s) Containing Crownophanes

4.1
Fluorenone and Stilbene Ring(s) Containing Crownophanes

Crownophanes **38** containing of fluorenone and stilbene fragment bridged by diethylene glycol and triethylene glycol unit were synthesized by a conventional method. The crystal structure and complexation behavior of these crownophanes were studied. They form much stronger complexes with dibenzylammonium hexafluorophosphate (log K_a value in CH_3CN: 3.92 ±

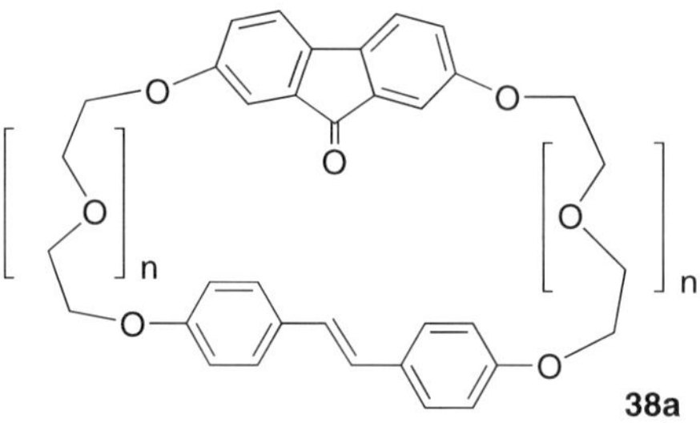

38a n = 1
b n = 2

0.06 for **38a** and 4.40 ± 0.05 for **38b**) than do the corresponding dibenzocrown ethers [41].

4.2
Anthracene Ring(s) Containing Crownophanes

Crownoanthracenophane **39** was designed to control their photoemission and photoreaction when Na$^+$ ions was incorporated in the crown rings. The photochemically interesting properties of **39** were reviewed [42].

Crownoanthracenophane **39** is known to be an excimer-type fluorosensor for Na$^+$ ions and to encapsulate electron-deficient species such as paraquat. Its bis-1,4-endoperoxide **40** is a tetraoxapaddlane. The molecular and crystal structures of **39** and **40** were reported [43].

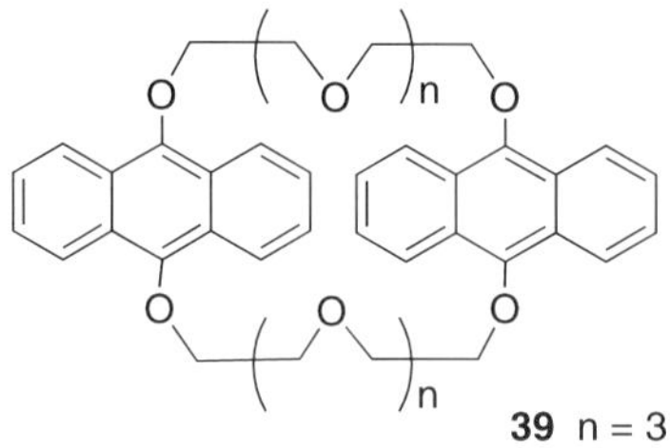

39 n = 3

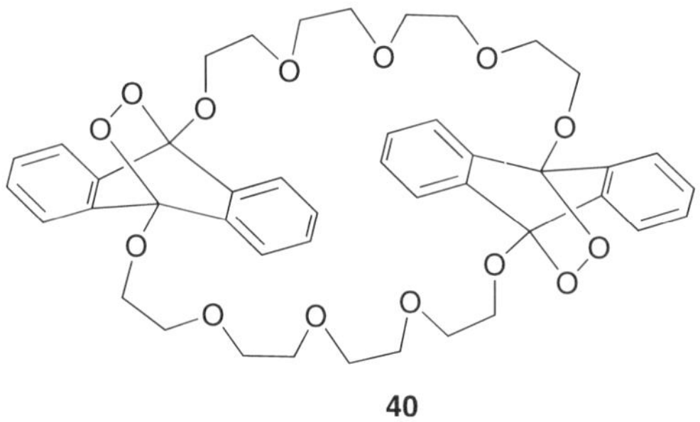

40

4.3
Pyrene Rings Containing Crownophanes

Crownopyrenophanes including **41** were prepared (method A), whose complexing behavior with anionic (naphthalene sulphonate derivative, DNA) and cationic aromatic (1,1'-dimethyl 4,4'-bipyridiniums) compounds was explored by UV and fluorescence spectroscopic analyses in MeOH and water-ethylene glycol mixed solvent. It was suggested that the binding affinities of the pyrenophanes for aromatic compounds were mainly governed by hydrophobic and/or π-stacking interaction [44].

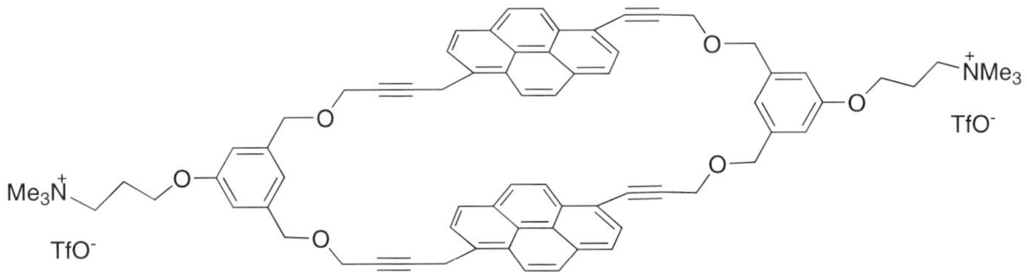

41

Similarly, water-soluble pyrenophanes having polycationic or amphiphilic side chains have been prepared to study hydrophobic and/or π-stacking interactions whose typical example is depicted in **42**. The hexaammonium-, bis(diazoniacrown)-, and tetrakis[octa(oxyethylene)] derivatives were soluble in pure water. The cationic pyrenophane **42** strongly recognized anionic arenes including nucleotides. The recognition ability for nucleotides by the bis(diazoniacrown)phane **42** depends on the number of phosphate moieties [45]. This is thought to become useful architecture for bioorganic field since phane **42** showed very high complexing ability toward ATP in water (the association constant is 1.0×10^6 M^{-1}).

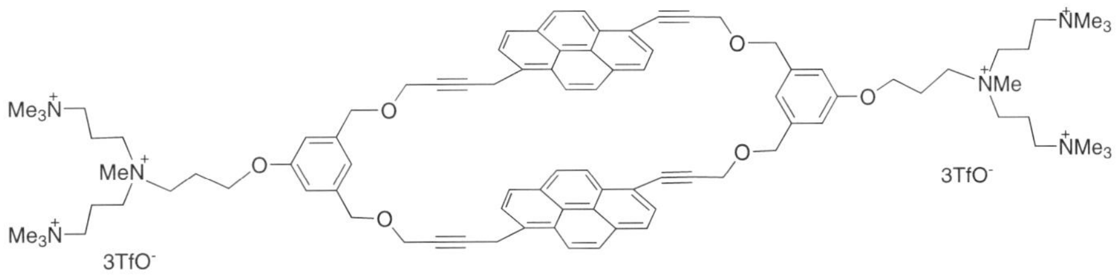

42

5
Heteroaromatic Ring Containing Crownophanes

5.1
Pyridine Ring(s) Containing Crownophanes

Inokuma et al. have prepared a new type of crownopyridinophanes **43** by method C. They were of *cis*-configuration with respect to the cyclobutane ring. According to ESI-MS analysis, compounds **43** formed the 1 : 1 complexes with Ag^+ cation. All compounds showed three sets of aromatic proton peaks, which were high-field shifted compared to those of compounds **44** ($\Delta\delta = 0.25$–0.39), indicating the phane structure having well-overlapped layer aromatic nuclei. In a liquid–liquid extraction, **43** showed the highest affinity toward Ag^+ cation among several heavy metal cations (Ag^+, Pb^{2+}, Cu^{2+}, Mn^{2+}, Zn^{2+}, Ni^{2+}, Co^{2+} and Fe^{3+}). In this series, **43b** possessing four ethereal oxygen atoms was found to show the highest Ag^+-affinity, according to the liquid–liquid extraction and determination of stability constant with the cation [46, 47].

Crownopyridinophanes **45** and **46** (prepared by method C) with three pyridine moieties exhibited high efficiency toward Ag^+. By comparing the high extractability and complexing stability constant for Ag^+ to those of the corresponding pyridinocrownophanes **47** and **48** and observing the ^{1}H NMR spectra in the presence of Ag$^+$, the ethereal oxygen atoms and the three nitrogen atoms were found efficiently and cooperatively to act as ligating sites [48]. In this way, the intramolecular photocycloaddition of styrene derivatives could be equally applied to vinylpyridine derivatives. This photocycloaddition was thought to be a promising method for more excellent extracting agent like bipyridine rings containing crownophanes described later because the crownopyridinophanes were efficient Ag$^+$-ligands as mentioned above.

49 R^1 = R^2 = H, X = CH$_2$SCH$_2$
50 R^1 = OMe, R^2 = Me, X = CH$_2$SCH$_2$
51 R^1 = R^2 = H, X = 2,6-Py
52 R^1 = OMe, R^2 = Me, X = 2,6-Py

53 X = CH$_2$SCH$_2$
54 X = 2,6-Py

55

The crownophanes possessing pyridine-nitrogen and sulfur atoms extracted Ag$^+$ with high efficiency. For example, crownophanes **49** and **51** extract Ag$^+$ 172 and 602 times more than Pb^{2+}, respectively [49].

5.2
Bipyridine Ring(s) Containing Crownophanes

Bipyridine **56** formed dinuclear double helicate ([Hg$_2$L$_2$Na$_2$]$^{6+}$) in the presence of Hg^{2+} and Na$^+$, whereas a mononuclear species is formed ([HgLBa]$^{4+}$) in the presence of Hg^{2+} and Ba^{2+} exclusively [50].

The ligand **57** formed a dinuclear double helicate with Cu$^+$ [Cu$_2$L$_2$]$^{2+}$, but upon addition of Ba^{2+} to the system a side-by-side species, {[Cu$_2$L$_2$Ba$_2$](ClO$_4$)$_4$(MeCN)$_4$}$^{2+}$, was formed both in solution and the solid state. In the presence of Na$^+$ both the helicates and side-by-side species, [Cu$_2$L$_2$Na$_2$]$^{4+}$, were formed in roughly equal amounts in solution [51].

Typical cyclophane structures having layered bipyridine nuclei have been dealt with in a few papers to the best of our knowledge, so that we prepared bipyridinocrownophanes **58a** and **58b** by method C [52]. As illustrated in Fig. 2, the solid-state structure of free ligand **58a** have layered aromatic nuclei. In the liquid–liquid extraction of heavy metal cations, **58a** and **58b** exhibited perfect selectively toward Ag$^+$ with high efficiency. It was found that the ethereal oxygen atoms and the four nitrogen atoms in **58a** and **58b** acted as ligating sites, according to the high extractability and com-

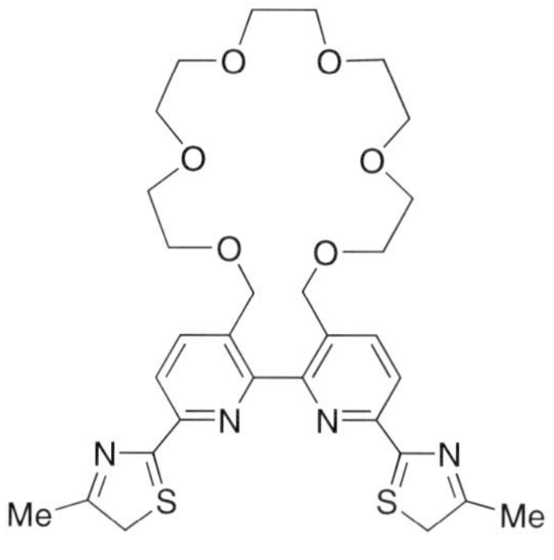

56

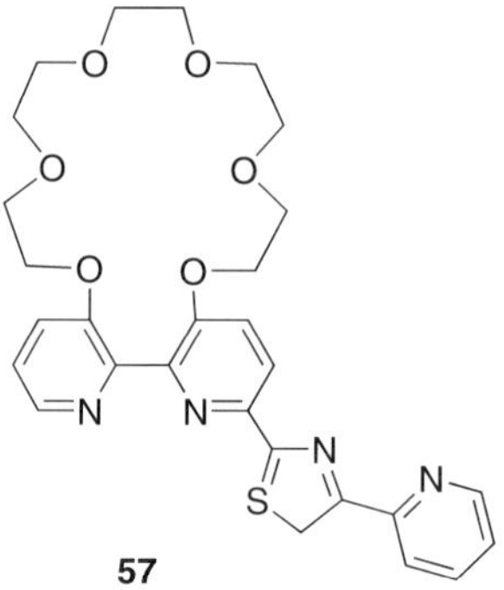

57

plexing stability constant for Ag⁺ compared to those of the corresponding pyridinocrownophanes **43c** and **43d**. ^{13}C NMR and ESI-MS analysis suggested that the crownophanes formed a 1 : 1 complexes with the Ag⁺ ion.)

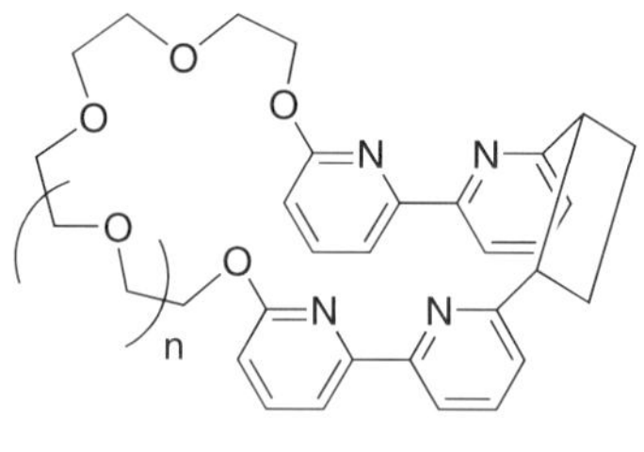

58a n=1 Y. 41%
b n=2 Y. 61%

5.3
Phenanethroline Ring(s) Containing Crownophanes

Recently, binding constants (log K) of ligand **59**-4H⁺ with a variety of di- and tricarboxylate anions were examined by ^{1}H NMR technique in D$_2$O

Syntheses and Properties of Crownophanes 61

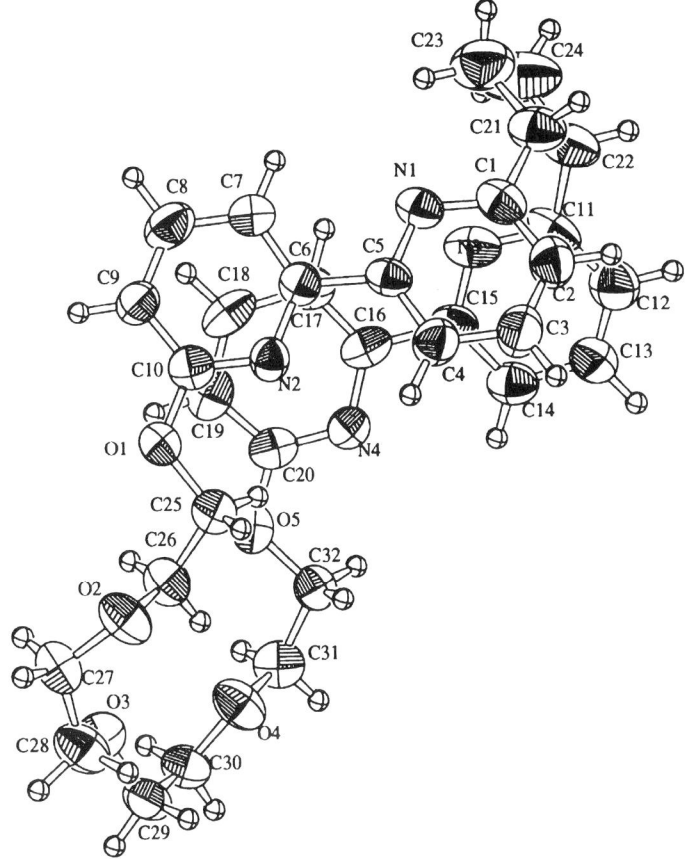

Fig. 2 ORTEP drawing of **58a** [52]

59

at 298 K. The order was 1,3,5-benzenetricarboxylate anion (> 5.0) > 1,3-benzenedicarboxylate anion (4.77) > 1,2-benzenedicarboxylate anion (3.23) > 1,3,5-cyclohexanetricarboxylate anion (3.08) > oxalate anion (2.59) [53].

6
Catenanes and Rotaxanes

In earlier works di(p-phenylene) crown such as di(p-phenylene)-34-crown-10 **60** was principally employed as a host molecule for alkali metal, alkaline earth metal and ammonium [54] and pyridinium cations including paraquat and diquat dications [55, 56]. In recent years, it has been exclusively used as donor components for rotaxanes or catenanes having secondary ammonium, bipyridinium rod or bipyridinium macrocycle [57–75].

60   **61**   **62**

A large number of catenanes and rotaxanes bearing 1,5-dinaphtho-38-crown-10, e.g. **61**, as a donor have been reported. As seen in recent papers, paraquat and diquat dication macrocycles or some rod derivatives as acceptor components are often utilized [76–85].

Some of them used diazapyrenium-based [86–88], TTF-based [89, 90], and diimide-based rods or diimide-based macrocycles including two kinds of diimides as acceptor parts [91–98], e.g., compounds **64–67**.

Rotaxanes bearing both 1,5-naphtho-38-crown-10 and TTF as donor parts have been reported [86]. 1,5-Naphtho-p-phenylene-36-crown-10 **62** was also employed as donor component of catenanes, e.g., compound **68** [99–101] and roraxanes [102, 103] having ammonium or bipyridinium cations as acceptor moieties.

Catenane **69** was obtained via ring-closing metathesis from starting material **70** in high overall yield (51%) from commercially available 1,10-phrenanthroline [104].

Syntheses and Properties of Crownophanes

63

64

65

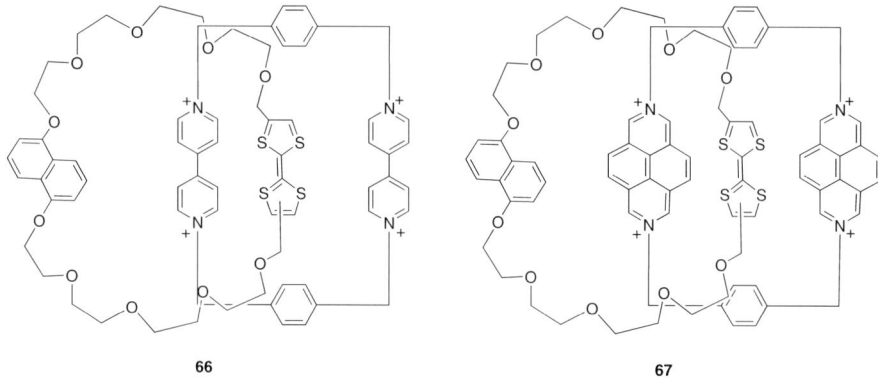

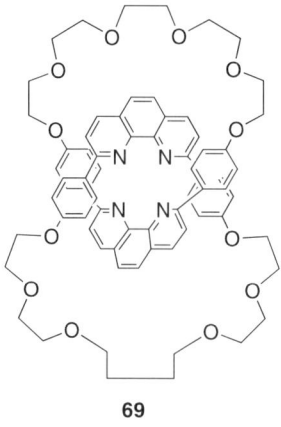

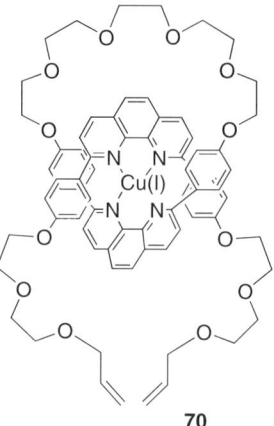

Hiratani et al. developed a new excellent methodology to make rotaxanes **71** via covalent bond formation. The rotaxanes **71** composed of crownophanes having two naphthol moieties as a rotor and an axle having diamide moieties were prepared via three steps: Claisen rearrangement (method B), diesterification, and aminolysis. The best yield of the rotaxane (56%) was recorded in the reaction of **72** with 9-aminomethylanthracene [105].

Rotaxanes **73** were prepared by treating macrocyclic monoesters **74** with $R_2(CH_2)_nNH_2$ in good yields. The association constant of rotaxane **73** with Li^+ was 4.2×10^4 M^{-1} [106].

Chiral rotaxanes **75–77** composed of the asymmetric crownophane incorporating two hydroxyl groups as a rotor moiety and asymmetric axis were effectively synthesized via covalent bond formation, i.e., tandem Claisen rearrangement as described above, esterification, and aminolysis [107].

The [1]Rotaxane **78** via covalent bond formation was firstly prepared by Hiratani et al. It was found that only Li^+ cation among alkali metal cations could change the chemical shift of the rotaxane in ^1H NMR spectrum. Astonishingly, only Li^+ ion drastically enhanced the fluorescence intensity due to energy transfer occurred perfectly from the naphthalene ring of the rotor to anthracene ring of the axle. This might make it a candidate for a Li^+ ion sensor [108].

The [3]Rotaxane **79** composed of two 25-membered crownophanes and one axle molecule having two anthryl end groups was successfully synthesized via the covalent bond formation followed by aminolysis. It made

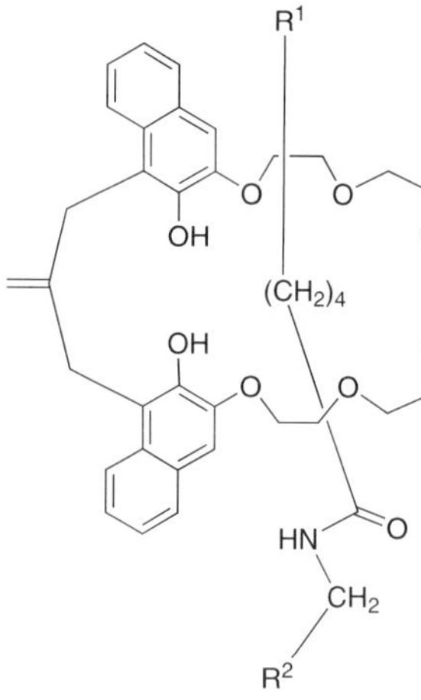
R1
OH
(CH2)4
OH
HN
CH2
R2

73

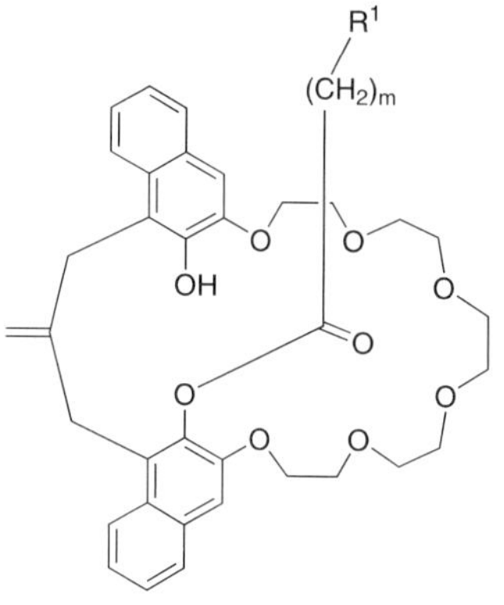
R1
(CH2)m
OH

74

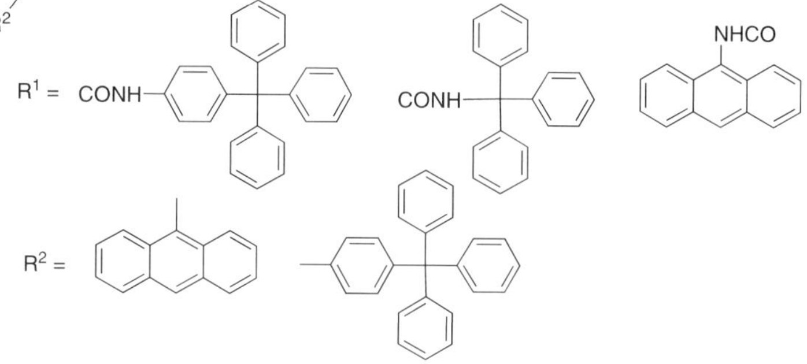
R1 =
CONH
CONH
NHCO
R2 =

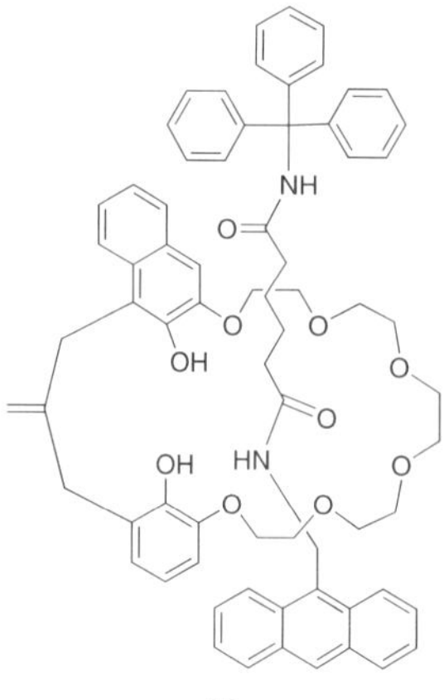
NH
OH
OH
HN

75

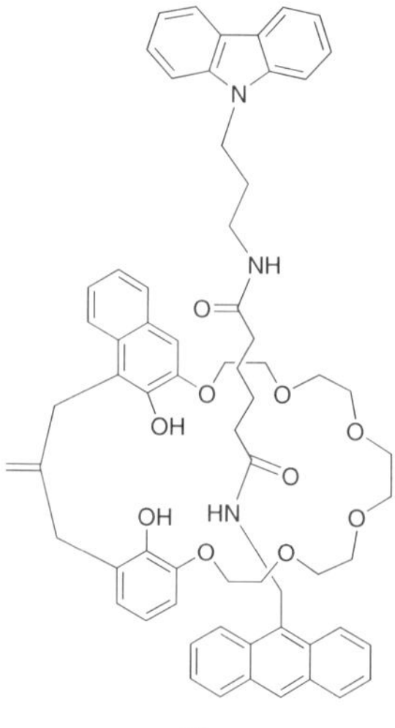
N
NH
OH
OH
HN

76

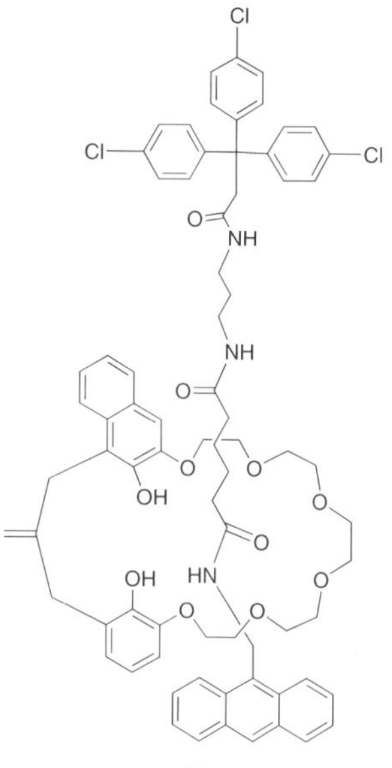
Cl
Cl
Cl
NH
NH
OH
OH
HN

77

a 1 : 2 complex with Li⁺ ion, whereas it incorporated Cs⁺ ion into the space between the two macrocycles as a 1 : 1 sandwich-type complex with the stability constant of 6.3×10^5 M^{-1} [39, 109]. The excellent yields of rotaxanes by the covalent bond formation were recently rationalized by Hirose et al. [110]. Thus, the presence of the crown ether ring significantly accelerates the aminolysis, presumably because it stabilizes the tetrahedral intermediates by hydrogen bonding [111]. As mentioned above, Hiratani's research group has elegantly prepared [3]rotaxane **79** via tandem Claisen rearrangement (method B).

7
Concluding Remarks

Simple crown ethers may be being examined extensively, but crownophanes will become more important as components in the supramolecular chemistry, because they can provide much more sites for modification by simple and facile reactions as mentioned above. For example, rotaxanes and catenanes attract much attention in contemporary research fields, whose building blocks are definitely some of the crownophanes.

References

1. Pedersen CJ (1967) J Am Chem Soc 89:7017
2. Sutherland IO (1990) In: Gokel GW (ed) Advances in Supermolecular Chemistry, Chap 1. JAI Press, Greenwich, p 65
3. An H-Y, Bradshaw JS, Izatt RM, Yan Z (1994) Chem Rev 94:939
4. Gokel GW (1992) Chem Soc Rev 39:3
5. Inokuma S, Funaki T, Nishimura J (2002) In: Takemura H (ed) Cyclophanes chemistry for the 21th Century, Chap. 7. Research Signpost, Kerala, India, p 149
6. Inokuma S, Sakai S, Nishimura J (1994) Top Curr Chem 172:87
7. Inokuma S, Yamamoto T, Nishimura (1990) Tetrahedron Lett 31:97
8. Inokuma S, Sakai S, Yamamoto T, Nishimura J (1994) J Membr Sci 97:175
9. Inokuma S, Gao S-R, Nishimura J (1995) Chem Lett, p 689
10. Inokuma S, Katoh R, Yamamoto T, Nishimura J (1991) Chem Lett, p 1751
11. Inokuma S, Saka S, Katoh R, Yasuda T, Nishimura J (1993) Nippon Kagaku Kaishi, p 1148
12. Inokuma S, Sakai S, Katoh R, Nishimura J (1994) Bull Chem Soc Jpn 67:1462
13. Inokuma S, Yasuda T, Araki S, Sakai S, Nishimura J (1994) Chem Lett, p 201
14. Inokuma S, Kimura K, Funaki T, Nishimura J (2001) Heterocycles 55:447
15. Inokuma S, Funaki T, Kondo S, Hara D, Nishimura J (2004) Heterocycles 63:333
16. Mateo C, Pérez-Melero C, Peláez R, Medardé M (2005) Tetrahedron Lett 46:7055
17. Pigge FC, Ghasedi F, Rath NP (2002) J Org Chem 67:4547
18. Pigge FC, Ghasedi F, Schitt AV, Dighe MK, Rath NP (2005) Tetrahedron 61:5363
19. Hiratani K, Uzawa H, Kasuga K, Kambayashi H (1997) Tetrahedron Lett 38:8993
20. Hiratani K, Uzawa H, Kasuga K, Kambayashi H (1997) J Am Chem Soc 119:12677
21. Inokuma S, Kobayashi A, Katoh R, Yasuda T, Nishimura J (1995) Heterocycles 40:401
22. Inokuma S, Funaki T, Tanakajima M, Isawa K, Nishimura J (2001) Heterocycles 55:1635
23. Inokuma S, Kimura K, Nishimura J (2001) J Inclu Phenom Macro Chem 39:30
24. Inokuma S, Kimura K, Nishimura J (1998) Chem Lett, p 287
25. Siebert JW, Hundt GR, Sargent AL, Lynch V (2005) Tetrahedron 61:12350
26. Siebert JW, Forshee PB, Lynch V (2005) Inorg Chem 44:8602
27. Kim H-Y, Lee W-J, Kang H-M, Cho C-G (2007) Org Lett 16:3185
28. Inokuma S, Takezawa M, Satoh H, Nakamura Y, Sasaki T, Nisimura J (1998) J Org Chem 63:5791
29. Inokuma S, Sakaizawa T, Funaki T, Yonekura T, Satoh H, Kondo S, Nakamura Y, Nishimura J (2003) Tetrahedron 59:8183
30. Xu J, Lai Y-H (2002) Org Lett 4:3211
31. Xu J, Lai Y-H (2002) Tetrahedron Lett 43:9199
32. Xu J, Lai Y-H, Wang W (2003) Org Lett 5:2781
33. Mertens IJA, Wegh R, Jenneskens LW, Vliestra EJ, van der K-van Hoof A, Zwikker JW, Cleiji TJ, Wilberth WJJ, Veldman N, Spek AL (1998) J Chem Soc Perkin Trans 2:725

34. Houjou H, Ogiwara T, Nagawa Y, Hiratani K (2001) J Inclu Phenom Macro Chem 39:347
35. Tokuhisa H, Nagawa Y, Uzawa H, Hiratani K (1999) Tetrahedron Lett 40:8007
36. Nagawa Y, Fukazawa N, Suga J, Horn M, Tokuhisa H, Hiratani K, Watanabe K (2000) Tetrahedron Lett 41:9261
37. Hiratani K, Sakamoto N, Kameta N, Karikomi M, Nagawa Y (2004) Chem Commun, p 1474
38. Hiratani K, Nagawa Y, Kanasato M (2005) Japan Kokai Tokkyo Koho, p 9 (CODEN:JKXXAF JP 2005075773 A 20050324)
39. Nagawa Y, Hiratani K, Koyama E, Kanesato M (2005) Japan Kokai Tokkyo Koho, p 12 (CODEN:JKXXAF JP 2005022991 A 20050127)
40. Naher S, Hiratani K, Ito S (2006) J Inclu Phenom Macro Chem 55:151
41. Lukyanenko NG, Kirichenko TI, Lyapunov AY, Kulygina CY, Lyapunov AY, Simonov YA, Fonari MS, Botoshansky MM (2004) Tetrahedron Lett 45:2927
42. Desvergne J-P, Perez-Inestrosa HE, Bouas-Lauarent H, Jonusauskas G, Oberle J, Rulliere C (2001) In: Valeur B, Brochon J-C (eds) Springer Series on Fluorescence 1 (New Trends in Fluorescence Spectroscopy). Springer 157
43. Marsa P, Guinand G, Hinschberger J, Desvergne J-P, Bouas-Lauarent H (2004) Aust J Chem 57:1085
44. Inouye M, Fujimoto K, Furusyo M, Nakazumi H (1999) J Am Chem Soc 121:1452
45. Abe H, Mawatari Y, Teraoka H, Fujimoto K, Inouye M (2004) J Org Chem 69:495
46. Funaki T, Inokuma S, Ide H, Yonekura T, Nakamura Y, Nishimura J (2004) Tetrahedron Lett 45:2393
47. Inokuma S, Ide H, Yonekura T, Funaki T, Kondo S, Shiobara S, Yoshihara T, Tobita S, Nishimura J (2005) J Org Chem 70:1698
48. Inokuma S, Yatsuzuka T, Ohtsuki S, Hino S, Nishimura J (2007) Tetrahedron 63:5088
49. Kumar S, Hundal M, Maninder S, Hundal G, Singh P, Bhalla V, Singh H (1998) J Chem Soc Perkin Trans 2:925
50. Baylies CJ, Harding LP, Jeffery JC, Riis-Johannessen T, Rice CR (2004) Angew Chem Int Ed 43:4515
51. Bokolinis G, Riis-Johannessen T, Harding LP, Jeffery JC, McLay N, Rice CR (2006) Chem Commun, p 1980
52. Inokuma S, Kurakami M, Otsuki S, Shirakawa T, Kondo S, Nakamura Y, Nishimura J (2006) Tetrahedron 62:10005
53. Cruz C, Delgado R, Drew MGB, Félix V (2007) J Org Chem 72:4023
54. Perez-Inestrosa E, Desvergne J-P, Bouas-Laurent H, Rayez J-C, Rayez M-T, Cotrait M, Marsau P (2002) Eur J Org Chem 331
55. Allwood BL, Spencer N, Shahriari-Zavareh H, Stoddard JF, Williams DJ (1987) J Chem Soc Chem Commun, p 1061
56. Ashton PR, Slawin AMZ, Spencer N, Stoddard JF, David DJ (1987) J Chem Soc Chem Commun, p 1066
57. Chas M, Blanco V, Peinador C, Quintela JM (2007) Org Lett 9:675
58. Chas M, Pia E, Toba R, Peinador C, Quintela JM (2006) Inorg Chem 45:6117
59. Halterman RL, David E, Pan X, Ha DB, Frow M, Haessig K (2006) Org Lett 8:2119
60. Nikitin K, Fizmaurice D (2005) J Am Chem Soc 127:8067
61. Orita A, Okano J, Tawa Y, Jiang L, Otera J (2004) Angew Chem Int Ed 43:3724
62. Feng D-J, Wang X-Z, Jiang X-K, Li Z-T (2004) Tetrahedron 60:6137
63. Jiang L, Okano J, Orita A, Otera J (2004) Angew Chem Int Ed 43:2121
64. Vibnon SA, Wong J, Tseng H-R, Stoddard JF (2004) Org Lett 6:1095
65. Long B, Nikitin K, Fizmaurice D (2003) J Am Chem Soc 125:15490
66. Huang F, Jones JW, Slebodnick C, Gibson HW (2003) J Am Chem Soc 125:14458

67. Benniston AC, Davies M, Harriman A, Sams C (2003) J Phys Chem A 107:4669
68. Nikitin, K Long B, Fizmaurice D (2003) Chem Commun, p 282
69. Amirsakis DG, Garcia-Garibay MA, Rowan SJ, Stoddart JF, White AJP, Williams DJ (2001) Angew Chem Int Ed 40:4256
70. Ashton PR, Baldoni V, Balzani V, Credi A, Hoffmann HDA, Martínez-Díaz M-V, Raymo FM, Stoddart JF, Venturi M (2001) Chem Eur J 7:3482
71. Ashton PR, Brown CL, Cao J, Lee J-Y, Simon SP, Raymo FM, Stoddard JF, White AJP, Williams DJ (2001) Eur J Org Chem 2001:957
72. Ashton PR, Becher J, Fyfe MCT, Nielsen MB, Stoddard JF, White AJP, Williams DJ (2001) Tetrahedron 57:947
73. D'Acerno C, Doddi G, Ercolani G, Mencarelli P (2000) Chem Eur J 6:3540
74. Wong EW, Collier CP, Behloradsky M, Raymo FM, Stoddard JF, Heath JR (2000) J Am Chem Soc 122:5831
75. Balzani V, Credi A, Langford S, Raymo FM, Stoddard JF, Venturi M (2000) J Am Chem Soc 122:3542
76. Mezei G, Kampf JW, Pecoraro VL (2007) New J Chem 31:439
77. Chas M, Blanco V, Peinador C, Quintela JM (2007) Org Lett 9:675
78. Mendes PM, Lu W, Tseng H-R, Shinder S, Iijima T, Miyaji M, Knobler CM, Stoddard JF (2006) J Phys Chem B 110:3845
79. Gunter MJ, Merican Z (2005) Supramol Chem 17:521
80. Flood AH, Peters AJ, Scott SA, David W, Tseng H-R, Kang S, James JR, Stoddard JF (2004) Chem Eur J 10:6558
81. Ashton PR, Baldoni V, Balzani V, Credi A, Hoffmann HDA, Martinez-Diaz M-V, Raymo FM, Stoddard JF (2001) Chem Eur J 7:3482
82. Cabezon B, Cao J, Raymo FM, Stoddard JF, White AJP (2000) Chem Eur J 6:2262
83. Cabezon B, Cao J, Raymo FM, Stoddard JF, White AJP, Williams DJ (2000) Angew Chem Int Ed 39:148
84. Zhang Q, Hamilton DG, Feeder N, Simon SJ, Goodman JM, Sanders JKM (1999) New J Chem 23:897
85. Amabilino DB, Ashton PR, Balzani V, Boyd SE, Credi A, Lee JY, Menzer S, Stoddard JF, Venturi M, Williams DJ (1998) J Am Chem Soc 120:4295
86. Flood AH, Peters AJ, Vignon SA, Steuerman DW, Tseng H-R, Kang S, Heath JR, Stoddard JF (2004) Chem Eur J 10:6558
87. Ashton PR, Boyd SE, Bribdle A, Langford SJ, Menzer S, Perez-Garcia L, Preece JA, Raymoisco M, Spencer N, Stoddard JF, White AJP, William DJ (1999) New J Chem 23:587
88. Ashton PR, Ballardini R, Balzani V, Constable EC, Credi A, Kocian O, Langford SJ, Preece JA, Prodi L, Schofield ER, Spencer N, Stoddard JF, Wenger S (1998) Chem Eur J 4:2413
89. Bryce MR, Cooke G, Florence MA, John P, Perepichka DF, Dmitrili F, Polwart N, Rotello VM, Stoddard JF (2003) J Mater Chem 13:2111
90. Ashton PR, Balzani V, Becher J, Credi A, Fyfe MCT, Mattersteig G, Menzer S, Nielsen MB, Raymo FM, Stoddard JF, Venturi M, Williams DJ (1999) J Am Chem Soc 121:3951
91. Vignon SA, Jarrosson T, Iijima T, Tseng H-R, Sanders JKM, Stoddard JF (2004) J Am Chem Soc 126:9884
92. Iijima T, Sanders JKM, Marchioni F, Venturi M, Apostoli E, Balzani V, Stoddard JF (2004) Chem Eur J 10:6375
93. Kaiser G, Jarrosson T, Otto S, Ng Y-F, Bond AD, Sanders JKM (2004) Angew Chem Int Ed 43:1959

94. Nakamura Y, Minami S, Iizuka K, Nishimura J (2003) Angew Chem Int Ed 42:3158
95. Gunter MJ, Bampos N, Johnstone KD, Leremy KM (2001) New J Chem 25:166
96. Hamilton DG, Montalti M, Prodi L, Fontani M, Zanello P, Sanders JKM (2000) Chem Eur J 6:608
97. Hansen JM, Feeder N, Hamilton DG, Gunter MJ, Becher J, Sanders JKM (2000) Org Lett 2:449
98. Hamilton DG, Davies JE, Prodi L, Sanders JMK (1998) Chem Eur J 4:608
99. Tseng H-R, Vibnon SA, Celestre PC, Stoddard JF, White AJP, Williams DJ (2003) Chem Eur J 9:543
100. Raymo FM, Houk KN, Stoddard JF (1998) J Org Chem 63:6523
101. Amabilino DB, Ashton PR, Stoddard JF, White AJP, Williams DJ (1998) Chem Eur J 4:460
102. Rogez G, Ferrer RB, Credi A, Ballardini R, Gandolfi MT, Balzani V, Liu Y, Brain H, Stoddard JF (2007) J Am Chem Soc 129:4633
103. Asakawa M, Ashton PR, Ballardini R, Balzani V, Belohradsky M, Gandolfi MT, Kocian O, Prodi L, Raymo FM, Stoddard JF, Venturi M (1997) J Am Chem Soc 119:302
104. Weck M, Mohr B, Sauvage JP, Grubbs RH (1999) J Org Chem 64:5463
105. Hiratani K, Suga J, Nagawa Y, Houjou H, Tokuhisa H, Numata M, Watanabe K (2002) Tetrahedron Lett 43:5747
106. Nagawa Y, Hiratani K, Koyama E (2004) Japan Kokai Tokkyo Koho, p 7 (CODEN:JKXXAF JP 2004018402 A 20040122)
107. Kameta N, Hirtani K, Nagawa Y (2004) Chem Commun, p 466
108. Hiratani K, Kaneyama M, Nagawa Y, Koyama E, Kanesato M (2004) J Am Chem Soc 126:13568
109. Nagawa Y, Suga J, Hiratani K, Koyama E, Kanesato M (2005) Chem Commun, p 749
110. Hirose K, Nishihara K, Harada N, Nakamura Y, Masuda D, Araki M, Tobe Y (2007) Org Lett 16:2969
111. Hogan JC, Gandour RD (1992) J Org Chem 57:55

Azacalixarene: A New Class in the Calixarene Family

Hirohito Tsue (✉) · Koichi Ishibashi · Rui Tamura

Graduate School of Human and Environmental Studies, Kyoto University, Sakyo-ku, 606-8501 Kyoto, Japan
tsue@ger.mbox.media.kyoto-u.ac.jp

1	Introduction	74
2	Syntheses of Azacalixarenes	75
2.1	Single-step Synthesis	75
2.2	Non-convergent Stepwise Synthesis	77
2.3	Convergent Fragment Coupling Synthesis	77
3	Structural Investigations of Azacalixarenes	82
3.1	Conformations in the Solid State	82
3.1.1	Azacalix[3]arene	82
3.1.2	Azacalix[4]arene	82
3.1.3	Azacalix[5]arene	85
3.1.4	Azacalix[6]arene	85
3.1.5	Azacalix[8]arene	85
3.1.6	Azacalix[10]arene	87
3.2	Conformations in Solution	88
4	Inclusion Properties of Azacalixarenes	90
5	Concluding Remarks	95
	References	95

Abstract Calixarenes, together with crown ethers and cyclodextrins, play an important role in supramolecular chemistry. A variety of calixarene analogues involving heteroatoms as the bridging units have been reported because the substitution of the carbon bridges with heteroatoms can impart novel properties and functions to molecules. This is typified by, for example, thiacalixarenes. In recent years, nitrogen-bridged calixarene analogues have emerged as a new calixarene family. While the diversity is still limited as compared with carbon- and sulfur-bridged calixarenes, intriguing structure-property relationships based on the introduction of nitrogen atoms as the bridging units have been reported. This review summarizes the recent reports on the preparations, conformations, and inclusion properties of nitrogen-bridged calixarene analogues with a $[1_n]$metacyclophane skeleton.

Keywords Azacalixarene · Conformation · Crystal structure · Inclusion property · Macrocyclization

Abbreviations
dba   *trans,trans*-1,5-diphenyl-1,4-pentadien-3-one
DMA   *N,N*-dimethylacetamide

DMPSCl   dimethylphenylsilyl chloride
DPEphos   bis[2-(diphenylphosphino)phenyl] ether
dppp   1,3-bis(diphenylphophino)propane
Xantphos 9,9-dimethyl-4,5-bis(diphenylphosphino)xanthene

1
Introduction

"Calixarene" is the term coined for a series of macrocyclic phenol condensates connected with methylene bridges, and are playing a significant role in supramolecular chemistry together with crown ethers and cyclodextrins [1–4]. The chemistry of calixarenes has been extensively studied over the last three decades, and the accumulated diverse knowledge of their conformations, complexation abilities, and chemical modifications renders them further intriguing. Replacement of the carbon bridges by heteroatoms offers a broad option to create calixarene analogues with both chemically and physically peculiar properties [5–7]. This has been clearly exemplified by thiacalixarenes that exhibit interesting structure-property relationships reflecting the substitution of the methylene-bridges with sulfur atoms [6, 7]. A variety of calixarene analogues involving heteroatoms other than sulfur atoms as the bridging units have been reported [4, 5], however, the chemistry of such analogues is still under investigation due to their limited accessibility. In this review, we will focus on less common nitrogen-bridged calixarene analogues since they have recently emerged as a new calixarene family by breaking a long and complete silence that has persisted since the first paper was published in 1963 on the X-ray crystallographic analysis of **1** [8]. In the following sections, a more familiar name, "azacalix[n]arene", is used to indicate nitrogen-bridged calixarene analogues with a [1_n]metacyclophane skeleton, and the current knowledge of their preparations, conformations, and inclusion properties are summarized.

1

For clarity, the following letters in Table 1 will be added to compound numbers of azacalix[n]arenes to identify ring size unless otherwise noted. For example, the above azacalix[4]arene **1** is designated as **1b**.

Table 1 Letters attached to the compound numbers of azacalix[n]arenes to identify ring size

n	3	4	5	6	7	8	10
letter	a	b	c	d	e	f	g

2 Syntheses of Azacalixarenes

For the synthesis of azacalixarenes, three typical synthetic strategies have thus far been employed that are substantially identical to those established in the calixarene chemistry; i.e., (1) single-step synthesis, (2) non-convergent stepwise synthesis, and (3) convergent fragment coupling synthesis [1–3]. In the following subsections, the preparations of azacalixarenes are described from these three strategic points of view.

2.1 Single-step Synthesis

A single-step procedure is the most efficient approach to azacalixarenes consisting of one single type of aromatic unit, as in the case of carbon-bridged calixarenes. However, a significant difference lies between the syntheses of calixarenes and azacalixarenes. Base-induced one-step procedures were established as standard protocols for preparing *p-tert*-butylcalix[n]arenes (n = 4, 6, and 8) [9–11], whereas palladium-catalyzed aryl amination reactions called "Buchwald-Hartwig reactions" [12–15], are often exploited as a key reaction for synthesizing azacalixarenes. As shown in Scheme 1, the Buchwald-Hartwig aryl amination reaction proceeds through the catalytic cycles involving (1) oxidative addition of aromatic halide to a palladium(0) catalyst, (2) coordination of aromatic amine to the palladium(II) species, (3) deprotonation of the coordinated NH group by a base, and (4) the final reductive elimination of N,N-diarylamine as a coupled product.

The first application of the single-step procedure to the synthesis of azacalixarenes was launched by Ito et al., who prepared a series of azacalix[n]arenes **4a–f** with different ring sizes (n = 3 to 8) [16, 17]. As shown in Scheme 2, starting from the N-methylation of 3-bromoaniline (**2**) in 3 steps, the resultant 3-bromo-N-methylaniline (**3**) was subjected to the Buchwald-Hartwig aryl amination reaction to yield azacalixarenes **4a–f**, which were isolated by medium-pressure liquid chromatography.

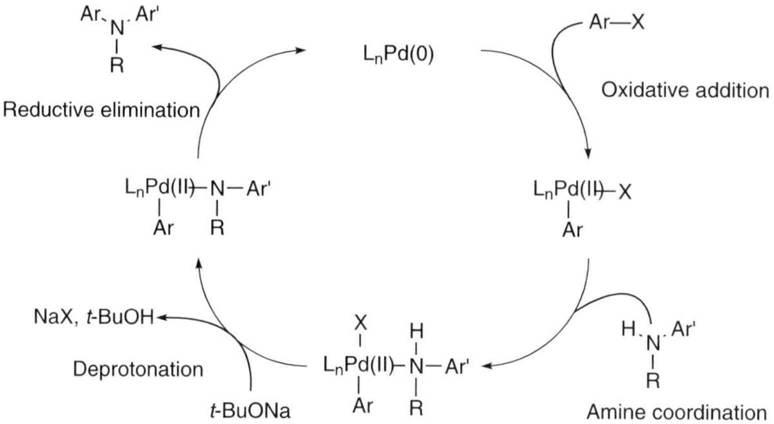

Scheme 1 Catalytic cycle of Buchwald-Hartwig aryl amination reaction of aromatic halide with N-alkylaniline [12]

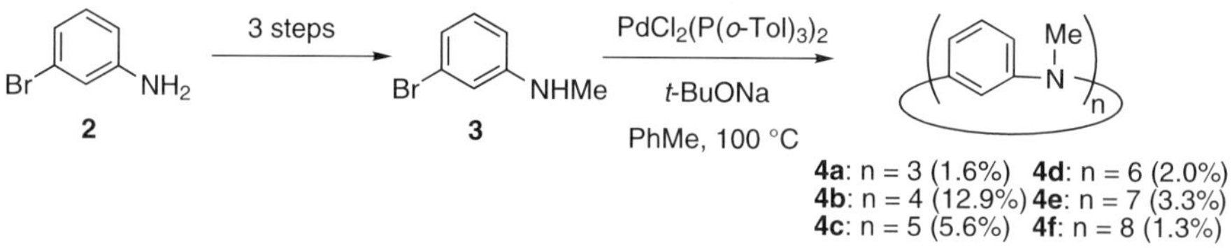

4a: n = 3 (1.6%) **4d**: n = 6 (2.0%)
4b: n = 4 (12.9%) **4e**: n = 7 (3.3%)
4c: n = 5 (5.6%) **4f**: n = 8 (1.3%)

Scheme 2 Single-step approach to azacalixarenes [16, 17]

Another application of the single-step procedure was made by Miyazaki et al., who utilized the Buchwald-Hartwig aryl amination reaction for the preparation of azacalix[6]pyridine **7d** [18], in which all the benzene rings of the above azacalix[6]arene **4d** were replaced by pyridine rings (Scheme 3). Azacalix[n]pyridines **8a–f** (n = 3–8) with tolyl groups on the nitrogen bridges

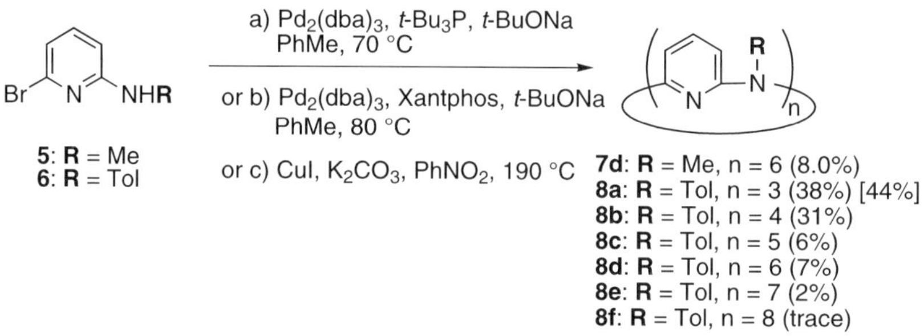

5: R = Me
6: R = Tol

7d: R = Me, n = 6 (8.0%)
8a: R = Tol, n = 3 (38%) [44%]
8b: R = Tol, n = 4 (31%)
8c: R = Tol, n = 5 (6%)
8d: R = Tol, n = 6 (7%)
8e: R = Tol, n = 7 (2%)
8f: R = Tol, n = 8 (trace)

Scheme 3 Single-step synthesis of azacalixpyridines. Reaction conditions a, b, and c were used for preparing **7d**, **8a–f**, and **8a**, respectively. The number in square brackets indicates the reaction yield of **8a** prepared under condition c [18, 19]

were also produced in a similar manner by Suzuki et al. [19]. Azacalix-pyridines **7d**, **8a**, **8b**, **8d**, and **8f** were prepared not only by the single-step procedure, but also by the convergent fragment coupling procedure, as described in Sect. 2.3.

2.2
Non-convergent Stepwise Synthesis

Historically, the non-convergent stepwise strategy was first devised by Hayes et al. in 1956 in order to synthesize carbon-bridged calixarenes from linear phenol oligomers [20, 21]. Only one application has thus far been reported for the non-convergent synthesis of azacalixarene. Tsue et al. prepared exhaustively methylated azacalix[4]arene **13b** [22] by applying the Buchwald-Hartwig aryl amination reaction for the intramolecular cyclization of linear tetramer **12** which was prepared from monomers **9** and **10** in four steps (Scheme 4).

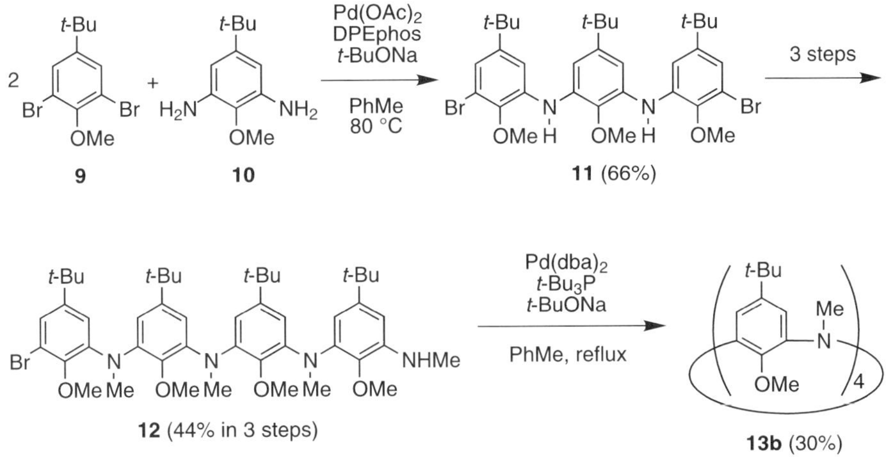

Scheme 4 Stepwise approach to azacalix[4]arene **13b** [22]

In principle, the non-convergent stepwise strategy can provide a versatile synthetic approach to azacalixarenes with different substituents on the aromatic rings and the nitrogen bridges, although thus far, the application of this strategy has been limited to the previously mentioned example.

2.3
Convergent Fragment Coupling Synthesis

In 1979, Böhmer et al. introduced a convergent fragment coupling strategy in which two different molecules were coupled in the cyclization step in order to yield carbon-bridged calixarenes [23]. This strategy can also provide a flexible synthetic approach to azacalixarenes, as the non-convergent stepwise method does (Sect. 2.2). In fact, a wide variety of azacalixarenes even with different aromatic π-systems have been prepared by applying this strategy.

As mentioned in the Sect. 2.1, azacalixpyridines **7d**, **8a**, **8b**, **8d**, and **8f** were prepared by using convergent fragment coupling as an additional method [18, 19] The macrocycles were synthesized by the Buchwald-Hartwig aryl amination reaction of two component monomers **14** and **15** or by the copper(I)-catalyzed coupling of **16** and **17** and of **14** and **17**, as shown in Scheme 5. It is worth noting that the fragment coupling synthesis of **7d** gave a slightly higher yield of 10.1% as compared with 8.0% of the single step synthesis of the same compound (Scheme 3 in Section 2.1). A similar outcome was obtained for the copper(I)-catalyzed syntheses of azacalixpyridines **8a**, **8d**, and **8f**.

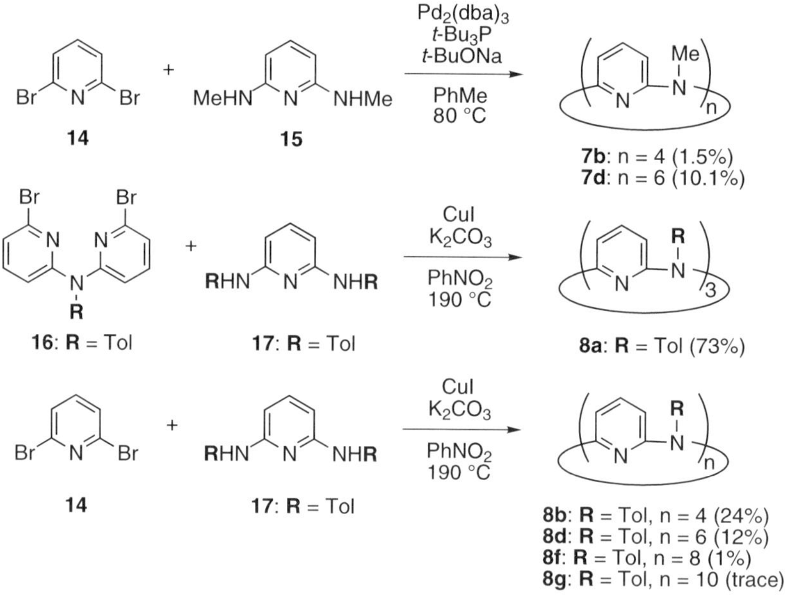

Scheme 5 Fragment-coupling approach to azacalixpyridines [18, 19]

Azacalixpyridines **7b** and **7f** were also independently prepared by Gong et al. [24], who used the convergent fragment coupling of linear pyridine trimer **18** and monomer **15** (Scheme 6). Trimer **18** was further coupled with pyridine dimer **21** in a similar manner to synthesize azacalixpyridines **7c** and **7g** [25]. Wang et al. studied the reaction conditions in detail and then succeeded in preparing "mixed" azacalixpyridines **20b** and **20f** designated as

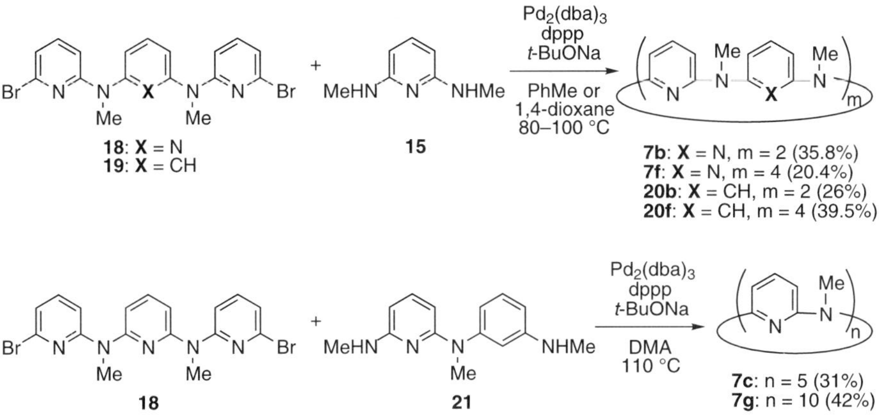

Scheme 6 Fragment-coupling synthesis of azacalixpyridines and azacalixarenepyridines [24–26]

azacalix[m]arene[m]pyridines [24, 26], in which benzene and pyridine rings were alternately linked by nitrogen bridges.

Another type of "mixed" azacalixarenes **25b**, **26b**, and **27b** were also reported by Wang et al. [27], who reported an efficient and convenient approach not only to these macrocycles, but also to oxygen-bridged analogues. As shown in Scheme 7, linear triazine trimer **22** was subjected to the aromatic nucleophilic substitution reaction with monomer **24** in order to yield azacalix[2]arene[2]triazine **25b**. The other analogues **26b** and **27b** were similarly prepared by the reactions of **22** and **15**, and of **23** and **15**. Larger homologues **28c** and **29d** were also produced in a similar manner by Graubaum et al. [28, 29].

Azacalixarenes **35b**, **35d**, **35f**, **36b**, and **36d**, which had no substituents on the nitrogen bridges, were successfully prepared by Fukushima et al. [30], as shown in Scheme 8. In contrast to the synthetic success of these macrocycles, they failed to prepare the parent azacalix[4]arene with no alkoxy groups. As indicated by these experimental facts, intramolecular NH···OR hydrogen bonding interactions lead to the folding and preorganization of the backbone of the uncyclized precursors (Fig. 1) and allow the intramolecular cyclization to efficiently afford azacalixarenes.

Azacalix[8]arene **38f** with partial NH bridges (Scheme 10) was prepared by a combination of the convergent fragment coupling method and a temporal N-silylation protocol [31], the latter of which was devised by Ishibashi et al. for suppressing an undesirable β-elimination reaction in the Buchwald-Hartwig aryl amination (Sect. 2.1). As shown in Scheme 9, β-elimination takes place as a side reaction when N-alkylaniline is used as a substrate [12, 32]. In the temporal N-silylation protocol, however, a silyl group without any hydrogens on the silicon atom is attached in situ onto an amino group in place of

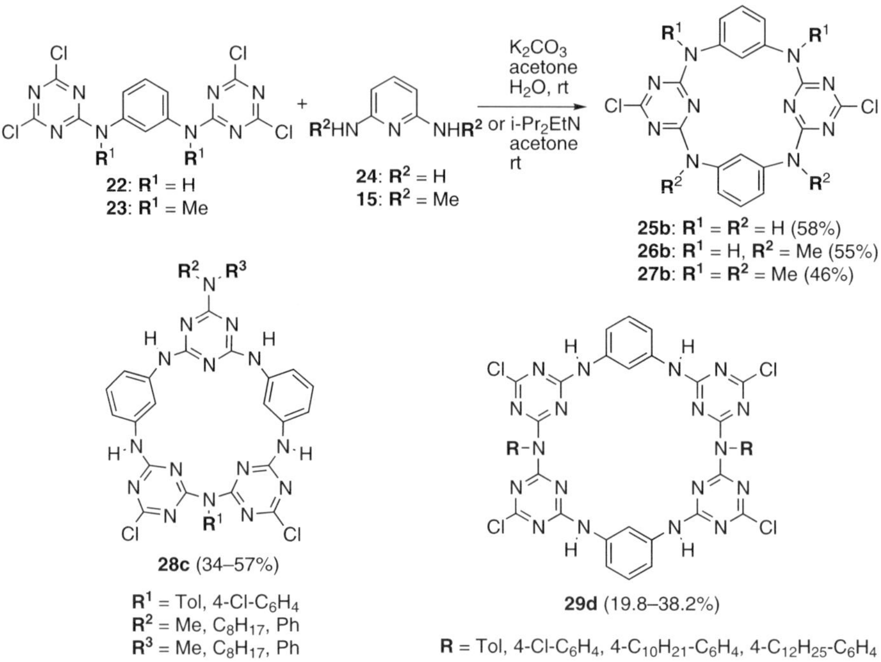
K2CO3
acetone
H2O, rt
or i-Pr2EtN
acetone
rt
22: R1 = H
23: R1 = Me
24: R2 = H
15: R2 = Me
25b: R1 = R2 = H (58%)
26b: R1 = H, R2 = Me (55%)
27b: R1 = R2 = Me (46%)
28c (34–57%)
R1 = Tol, 4-Cl-C6H4
R2 = Me, C8H17, Ph
R3 = Me, C8H17, Ph
29d (19.8–38.2%)
R = Tol, 4-Cl-C6H4, 4-C10H21-C6H4, 4-C12H25-C6H4

Scheme 7 Fragment-coupling synthesis of azacalixarenetriazines [27–29]

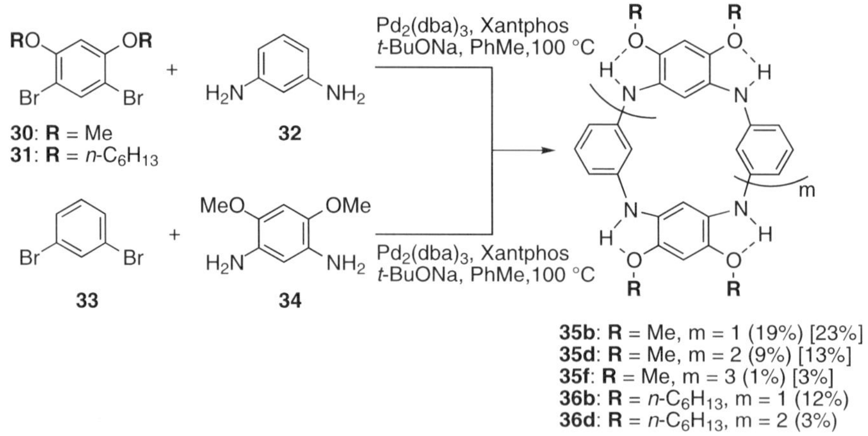
Pd2(dba)3, Xantphos
t-BuONa, PhMe,100 °C
30: R = Me
31: R = n-C6H13
32
33
34
35b: R = Me, m = 1 (19%) [23%]
35d: R = Me, m = 2 (9%) [13%]
35f: R = Me, m = 3 (1%) [3%]
36b: R = n-C6H13, m = 1 (12%)
36d: R = n-C6H13, m = 2 (3%)

Scheme 8 Fragment-coupling synthesis of azacalixarenes with NH bridges. Numbers in parentheses indicate the reaction yields in the fragment-coupling syntheses using monomers **30** and **32**, or **31** and **32** as substrates. Those in square brackets represent the yields in the reaction of **33** and **34** [30]

alkyl groups, thereby directing the favorable reductive elimination pathway to give the cross-coupling product. By applying this protocol, regioselectively *N*-methylated azacalix[8]arene **38f** was produced [31], as shown in Scheme 10.

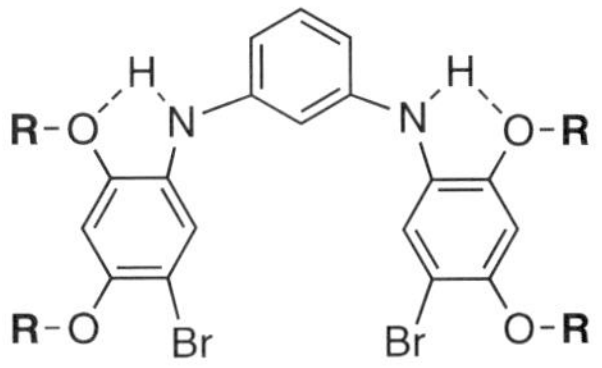

Fig. 1 Folding of the uncyclized precursor by intramolecular NH···OR hydrogen bonding interactions [30]

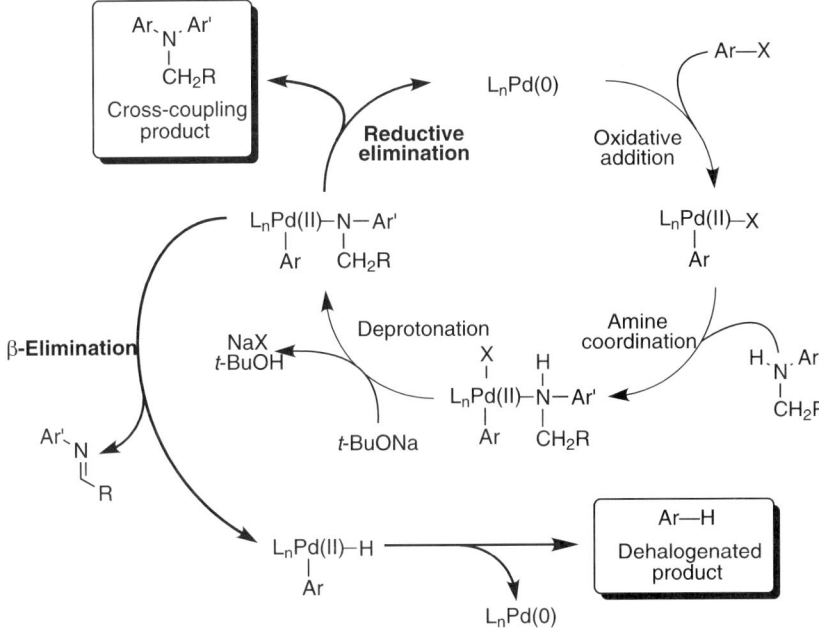

Scheme 9 β-Elimination reaction as a side reaction in the Buchwald-Hartwig aryl amination reaction [12]

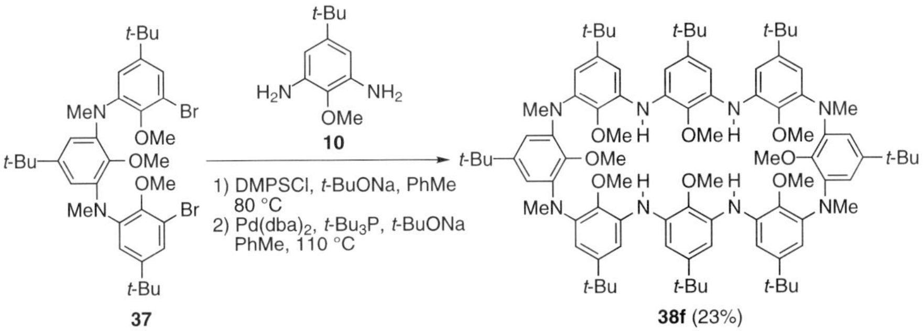

Scheme 10 Fragment-coupling synthesis of regioselectively N-methylated azacalix[8]arene 38f [31]

3
Structural Investigations of Azacalixarenes

3.1
Conformations in the Solid State

X-ray crystallographic analysis is the most powerful technique for providing information not only about molecular structures, but also about the key factors controlling their conformations. Until now, solid state structures of azacalix[n]arenes where n = 3, 4, 5, 6, 8, and 10 have been reported. In the following subsections, X-ray crystal structures of azacalixarenes will be described according to the ring sizes.

3.1.1
Azacalix[3]arene

Presently, the only known example is azacalix[3]pyridine **8a** [19]. In the solid state, it adopts a triangular shape with approximate C_s symmetry (Fig. 2). One pyridine ring is oriented to a different direction from the remaining two pyridine rings to avoid the electrostatic repulsion between the pyridine lone pairs in the cavity.

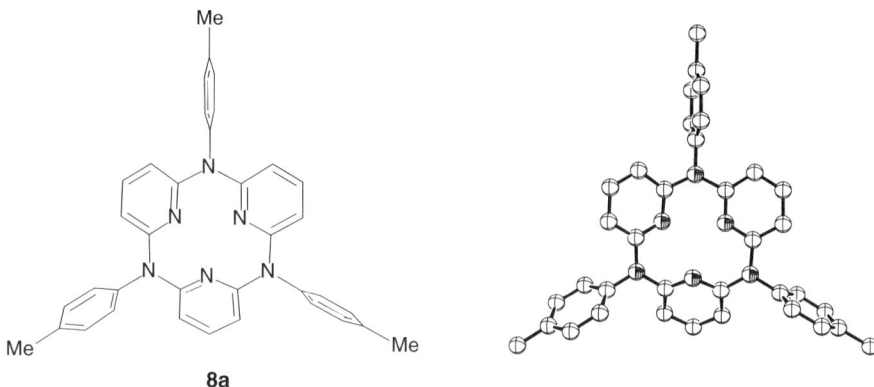

Fig. 2 X-ray crystal structure of azacalix[3]pyridine **8a** [19]

3.1.2
Azacalix[4]arene

The molecular structures of nine azacalix[4]arenes have been studied by X-ray crystallography. Intriguingly, all the reported azacalix[4]arenes exclusively adopt a 1,3-alternate conformation in the solid state, irrespective of the type of the aromatic π-systems involved in the macrocycles. Nonetheless, variations for the same 1,3-alternate conformation exist which can be classified, though arbitrarily, into three types on the basis of the molecular shapes.

The first type is a clip-like conformation observed in six azacalix[4]arenes. A typical example is azacalix[2]arene[2]triazine **25b** [27], which adopts a 1,3-alternate conformation with C_{2v} symmetry (Fig. 3). All the bridging nitrogen atoms conjugate with the triazine rings rather than with the benzene rings because of the electron-withdrawing nature of triazyl. Thus, the macrocycle can be viewed as a cyclic array comprised of two isolated benzene rings and two conjugated 2,6-bis(methylamino)triazine planar segments. As a result, a pair of opposite triazine rings is roughly coplanar, whereas a pair of benzene rings is almost parallel face-to-face, forming the clip-like conformation, as shown in Fig. 3. A similar conformation was also found in the other azacalix[2]arene[2]triazines **26b** [27] and **27b** [27], and azacalix[2]arene[2]pyridine **20b** [26]. Interestingly, azacalix[4]pyridines **7b** [24]

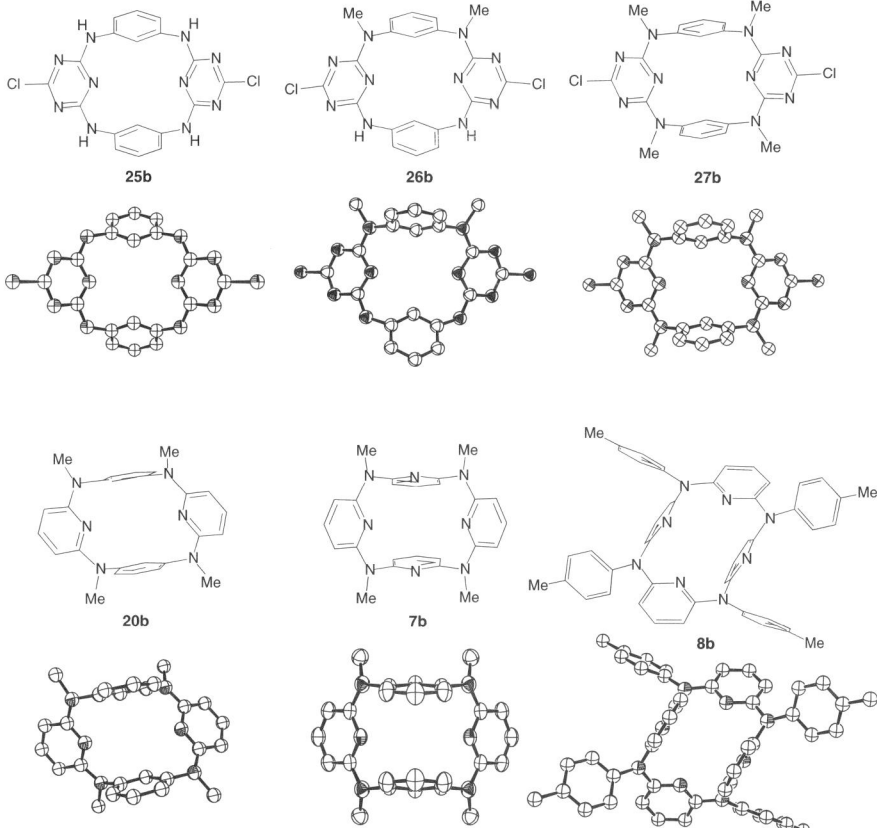

Fig. 3 X-ray crystal structures of azacalix[4]arenes with a clip-like 1,3-conformation [19, 24, 26, 27]

and **8b** [19] have an all-pyridine-based structure, but adopt a clip-like conformation.

The second type is a twisted 1,3-conformation observed in two azacalix[4]arenes, **4b** [16] and **35b** [30]. In the crystalline state, azacalix[4]arene **4b** adopts a twisted conformation with S_4 symmetry (Fig. 4). The benzene rings are alternately located upward and downward with respect to the molecular mean plane. Each bridging nitrogen conjugates with one of the neighboring benzene rings, and thus the macrocycle can be regarded as a cyclic array of four conjugated N-methylaniline planer units. Azacalix[4]arene **35b** also adopts a similar twisted 1,3-alternate conformation, though slightly distorted from S_4 symmetry because of the intramolecular NH···OMe hydrogen bonds.

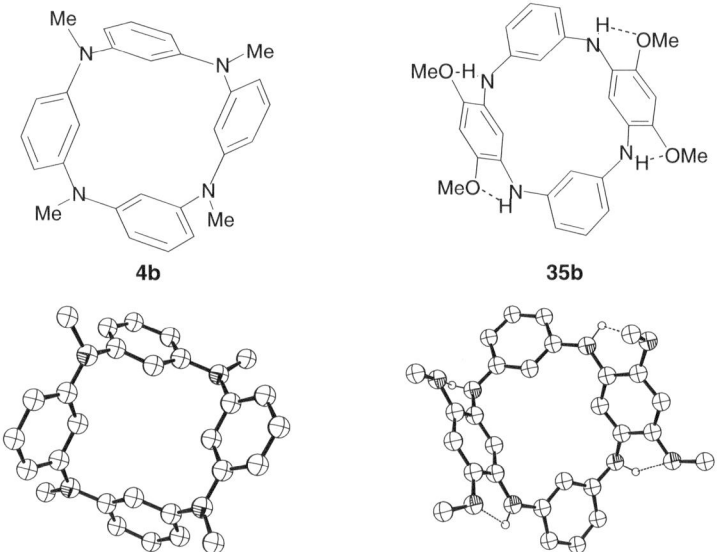

Fig. 4 X-ray crystal structures of azacalix[4]arenes with a twisted 1,3-conformation [16, 30]

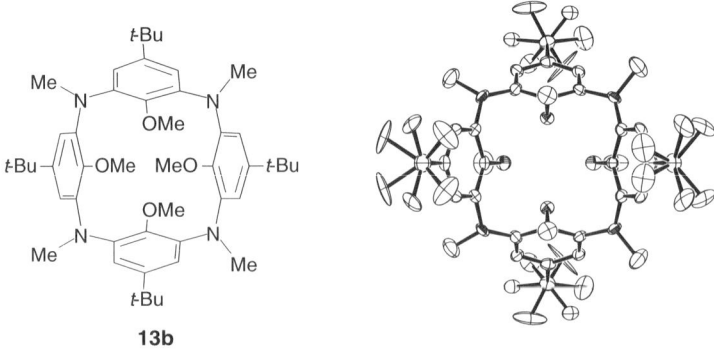

Fig. 5 X-ray crystal structure of azacalix[4]arene **13b** [22]

The third type is an ideal 1,3-alternate conformation with D_{2d} symmetry, and only one instance has been reported in azacalix[4]arene **13b** [22]. As shown in Fig. 5, it adopts a 1,3-alternate conformation with approximate D_{2d} symmetry in the solid state. Each bridging nitrogen atom adopts a planar sp^2 hybrid configuration and conjugates equally with two adjacent benzene rings. In addition, the methoxy groups are properly arranged so as to reduce steric hindrance as much as possible. As a result, a combination of electronic and steric effects render the macrocycle highly symmetrical.

3.1.3
Azacalix[5]arene

The only known example is azacalix[5]pyridine **7c** [25]. In the solid state, it adopts a distorted 1,3-alternate conformation, as shown in Fig. 6. Four pyridine rings are directed outward from the cavity, and the one remaining pyridine ring is inward. All the bridging nitrogen atoms adopt approximate sp^2 hybrid configurations and partially conjugate with both of their adjacent pyridine rings, as in the case of azacalix[4]arene **13b** (Sect. 3.1.2).

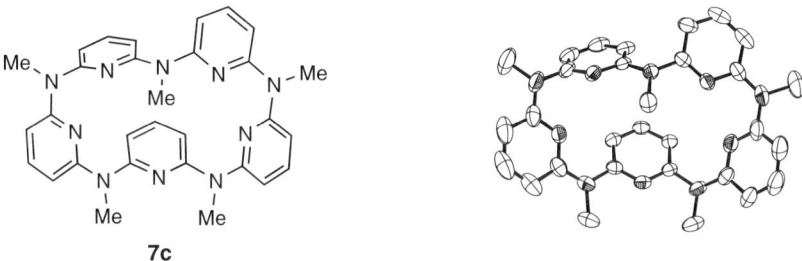

Fig. 6 X-ray crystal structure of azacalix[5]pyridine **7c** [25]

3.1.4
Azacalix[6]arene

Two azacalix[6]pyridines, **7d** [18] and **8d** [19], have thus far been investigated. In the solid state, both azacalix[6]pyridines adopt roughly triangular shapes, though the molecular geometries are different from each other to some extent, as shown in Fig. 7. The pyridine rings of **7d** and **8d** are arranged to minimize the electrostatic repulsion between the pyridine lone pairs, as in the case of azacalix[3]pyridine **8a** (Sect. 3.1.1).

3.1.5
Azacalix[8]arene

Three compounds have been studied involving azacalix[8]arene **38f** [31], azacalix[8]pyridine **7f** [24], and azacalix[4]arene[4]pyridine **20f** [26]. As

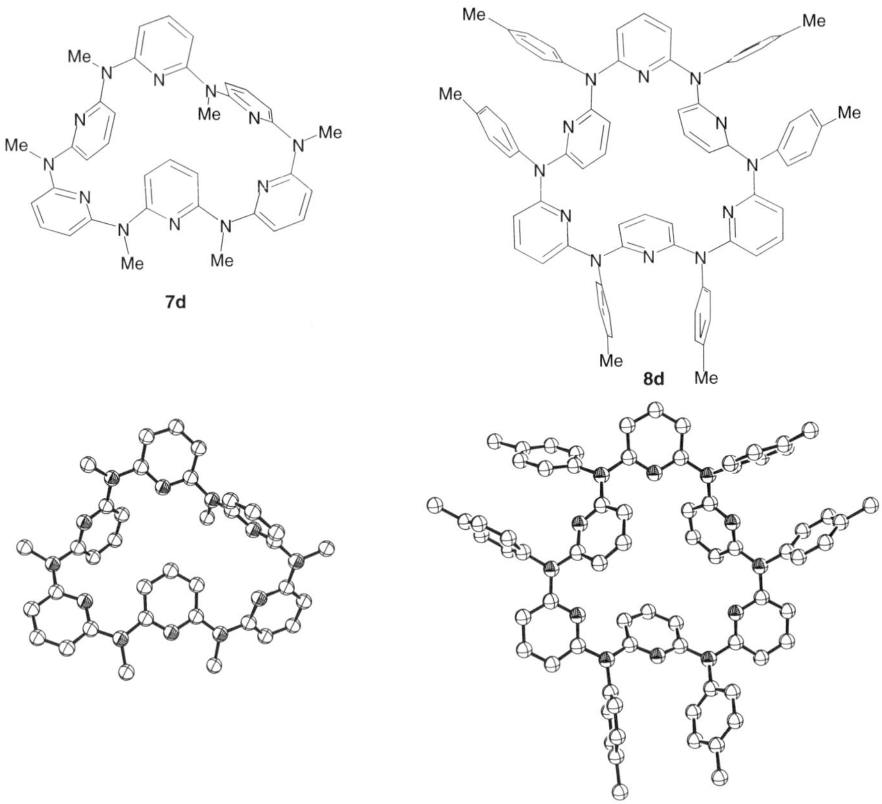

Fig. 7 X-ray crystal structures of azacalix[6]pyridines [18, 19]

shown in Fig. 8, azacalix[8]arene **38f** possesses a roughly ellipsoidal shape with C_2 symmetry in the crystalline state. All the benzene rings nearly lie on one plane, except for two opposite benzene rings that are tilted perpendicularly to the plane. The molecular geometry of **38f** is controlled by the intramolecular bifurcated MeO···NH···OMe hydrogen bonding interactions.

While azacalix[8]pyridine **7f** adopts a double-ended spoon-like conformation with C_i symmetry, azacalix[4]arene[4]pyridine **20f** possesses a pleated loop conformation with the same symmetry. In **7f**, four bridging nitrogen atoms partially conjugate with the two adjacent pyridine rings, whereas the remaining four bridging nitrogens are conjugated with one of the neighboring pyridine rings. In contrast, all the bridging nitrogen atoms of **20f** conjugate with pyridine rings rather than the benzene rings. The peculiar conjugation of the nitrogen bridges with aromatic rings shape the conformations of **7f** and **20f**, as in the case of azacalix[2]arene[2]triazine **25b** (Sect. 3.1.2).

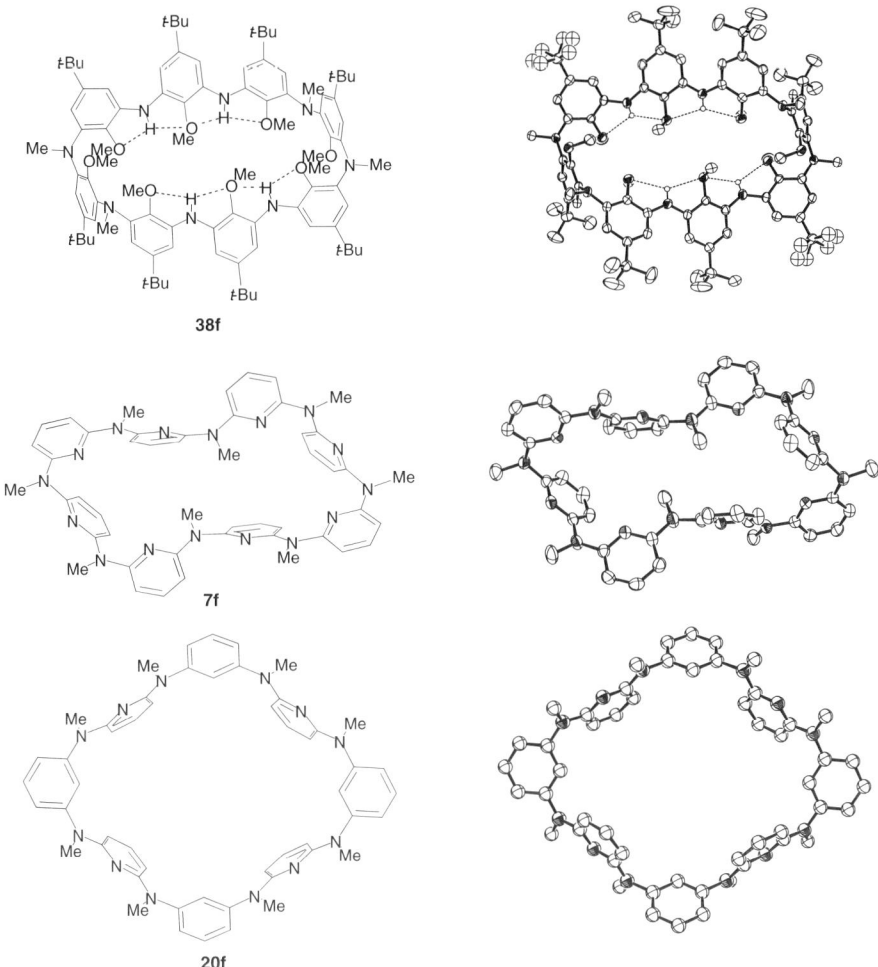

Fig. 8 X-ray crystal structures of azacalix[8]arenes [24, 26, 31]

3.1.6
Azacalix[10]arene

Only one example has been reported in azacalix[10]pyridine **7g** [25]. As shown in Fig. 9, it adopts a parallelogram shape with C_i symmetry. The four corner pyridine rings outwardly project from the cavity and are located on almost the same plane. The remaining six pyridine rings point to the oblique directions with respect to the plane. Each bridging nitrogen atom adopts an approximate sp^2 hybrid configuration and equally conjugates with two adjacent pyridine rings, as in the cases of azacalix[4]arene **13b** (Sect. 3.1.2) and azacalix[5]pyridine **7d** (Sect. 3.1.3).

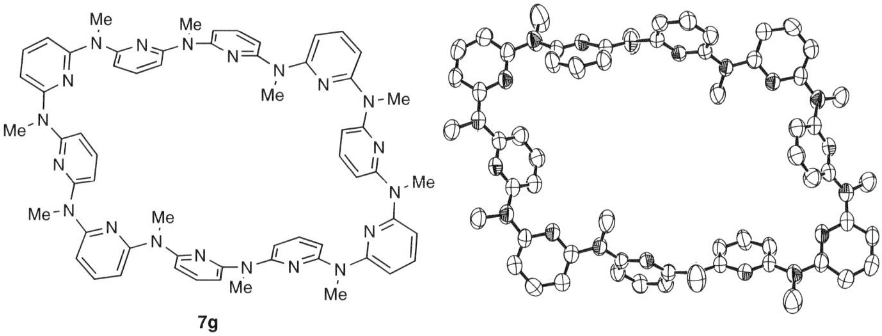

Fig. 9 X-ray crystal structure of azacalix[10]pyridine **7g** [25]

3.2
Conformations in Solution

All of the parent calixarenes with carbon bridges are conformationally flexible in solution [1, 3]. For instance, the cone conformation of calix[4]arene is transformed to its inverted cone conformation in solution at ambient temperature (Scheme 11).

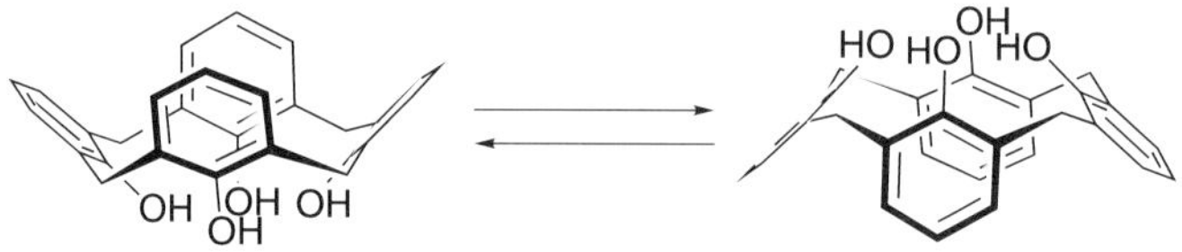

Scheme 11 Ring inversion of calix[4]arene [1, 3]

As in the case of carbon-bridged calixarenes, conformational behaviors of azacalixarenes in solution have been investigated by means of temperature dependent ^{1}H NMR spectroscopy. It was reported that azacalixarenes **4a–f**, **7b–d**, **7f**, **7g**, **8a**, **20b**, and **20f** were conformationally flexible in solution [17–19, 24, 25]. Gong et al. pointed out from their NMR studies of compounds **7b**, **7f**, **20b**, and **20f** that the lack of steric hindrance and intramolecular hydrogen bonds were responsible for their high conformational mobility in solution, as compared to carbon-bridged calixarenes [24].

Conversely, azacalixarenes with conformational inflexibility in solution were reported. Fukushima et al. [30] reported that the NH protons of azacalixarenes **35b**, **35d**, **35f**, **36b**, and **36d** were observed at δ 5.6–5.8 ppm as broad signals at room temperature and were sharpened at –50 °C. It was suggested from the NMR studies that intramolecular NH···OR hydrogen bonding interactions brought about their stiff conformations (cf. Fig. 4 in Sect. 3.1.2), which thus existed in a single conformation at the lower temperature.

Further interesting is azacalix[4]arene **13b** with a 1,3-alternate conformation, which has been demonstrated to be inflexible in solution [33]. Conformational behavior of **13b** in solution was examined by means of relaxation time measurements (Fig. 10). A much smaller longitudinal relaxation time of 1.03 s was observed for the aromatic protons of **13b**, as compared with 2.51 s reported for conformationally flexible *p-tert*-butylthiacalix[4]arene [34], demonstrating that the 1,3-conformation of **13b** was inflexible in solution. This result was further supported by two additional experimental facts. First, ^{1}H NMR spectra of **13b** were temperature independent [22, 33]. Second, the observed nuclear Overhauser effects were properly explained by considering a sole contribution of an inflexible 1,3-conformation of **13b** [22]. X-ray crystallographic analysis revealed that a small annulus of **13b** was responsible for the conformational immobilization by the small, but yet sufficiently bulky *O*-methyl groups [33], which were too small for carbon-bridged calix[4]arenes to keep their conformations in solution [1, 3, 35, 36].

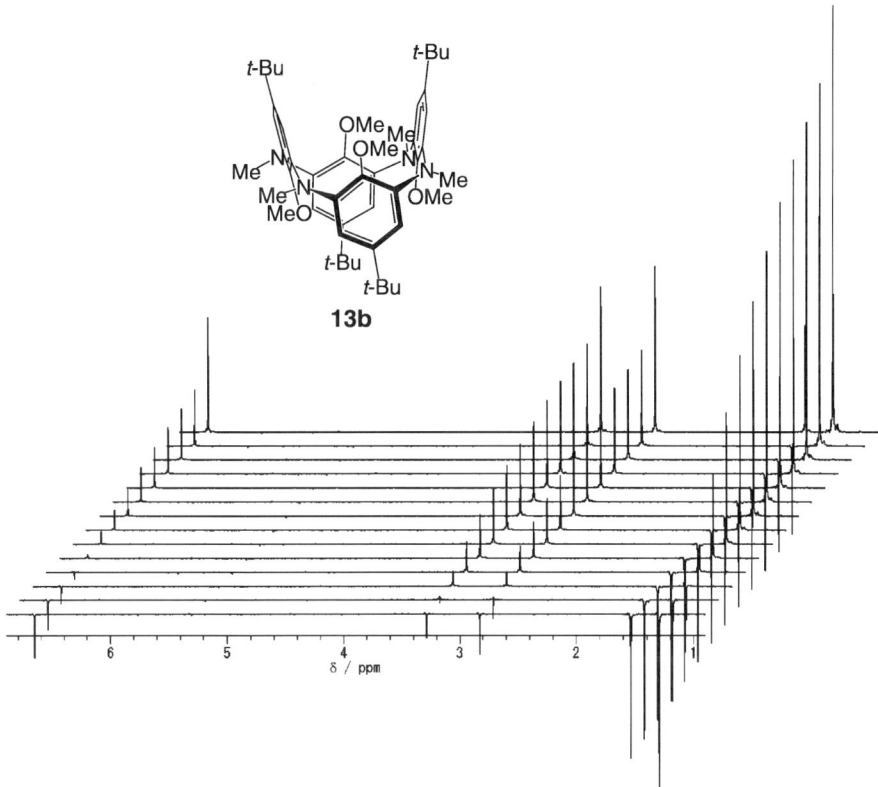

Fig. 10 ^{1}H NMR spectra from an inversion recovery experiment for azacalix[4]arene **13b** in CDCl$_3$ [33]

4
Inclusion Properties of Azacalixarenes

One of the most intriguing properties of carbon- and sulfur-bridged calixarenes is their ability to form complexes with a variety of organic and inorganic guest species. Inclusion properties of azacalixarenes remain relatively unexplored as of yet, appearing in only seven papers. In six of them, complexation behaviors of azacalix[n]pyridines **7b–d**, **7f**, **7g**, and **8a** as well as azacalix[m]arene[m]pyridines **20b** and **20f** were studied [18, 19, 24–26, 37]. In the remaining paper, azacalix[4]arene **13b** was investigated [33].

Tsue et al. investigated the host ability of azacalix[4]arene **13b** for alkali-metal cations [33]. Complexation behavior of **13b** for the cations was examined by ^{1}H NMR spectroscopy. As shown in Fig. 11, NMR signals of **13b** were drastically changed upon complexation with K^+ ion. Strong signal broadening was observed for the methoxy groups at 25 °C, and the resonances of the aromatic and *tert*-butyl protons were also broadened at –30 °C. Upon further decreasing the temperature to –60 °C, each of the broadened NMR signals was split into three, one of which corresponded to free **13b** and the remaining two were assigned to a 1 : 1 potassium complex [K **13b**]$^+$ depicted in Fig. 11. Na^+ and Li^+ ions were similarly examined, but no spectral changes were observed. As a result, azacalix[4]arene **13b** exhibited selective complexation for the K^+ ion on the basis of the inflexible 1,3-alternate conformation in solution (cf. Sect. 3.2).

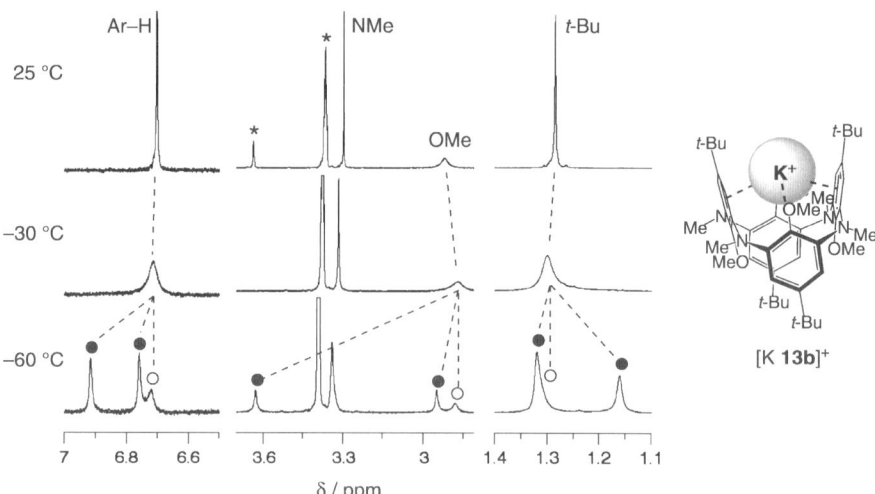

Fig. 11 Partial ^{1}H NMR spectra of azacalix[4]arene **13b** (0.50 mM) in the presence of potassium picrate (0.50 mM) in CDCl$_3$/CD$_3$OD (4 : 1, v/v). *Open* and *solid circles* represent the NMR resonances of free **13b** and those of potassium complex [K **13b**]$^+$, respectively. Peaks marked with an *asterisk* are due to solvent impurities [33]

Miyazaki et al. and Suzuki et al. reported that azacalixpyridines **8a** and **7b** were capable of forming copper(I) and zinc(II) complexes, respectively [18, 19]. X-ray crystallographic analysis of the copper complex of **8a** revealed three interesting structural features. First, as shown in Fig. 12, conformation of the complex is similar to that of free **8a** (Fig. 2 in Sect. 3.1.1). Second, the copper(I) ion is coordinated to two of the pyridine nitrogens. Third, the copper(I) ion is in a distorted trigonal-planar arrangement rather than a typical tetrahedral geometry. In the zinc complex of **7b**, the zinc(II) ion is embedded in the cavity in a slightly elongated octahedral coordination geometry with two water molecules as axial ligands (Fig. 12). The complex adopts an approximate S_4 conformation in which each pyridine ring is located alternately on both sides of the molecular mean plane. The zinc(II) ion is coordinated to four pyridine nitrogen atoms, and no bridging nitrogens participate in the coordination. Complexation behaviors of **8a** and **7b** with the relevant metal ions in solution were also studied by ^{1}H NMR spectroscopy, and these macrocycles as well as azacalix[6]pyridine **7d** were demonstrated forming metal complexes in solution [18, 19].

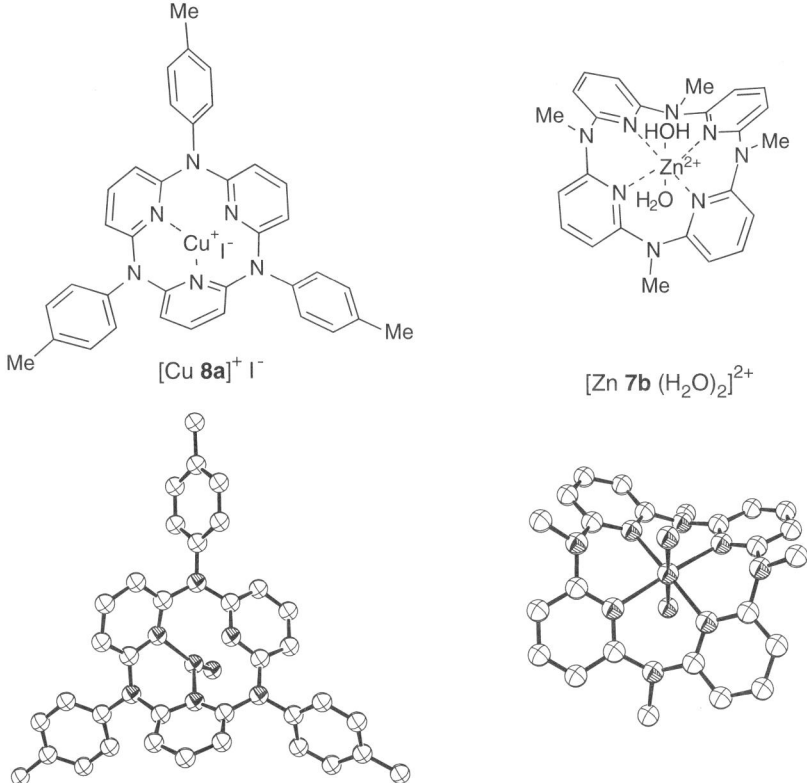

Fig. 12 X-ray crystal structures of metal complexes of azacalixpyridines [18, 19]

Kanbara et al. further studied the proton affinity of azacalix[3]pyridine **8a**, which behaved as a proton-sponge-like organic superbase [37]. To examine the basicity of **8a**, transportation experiments on **8a**, guanidine **39** (pK_{BH+} = 23.3 in MeCN), and proton sponge **40** (pK_{BH+} = 18.2–18.7 in MeCN) were carried out by ^{1}H NMR spectroscopy. As schematically represented in Scheme 12, both free **8a** and the protonated species [H **8a**]$^+$ were detected in the NMR experiments using 1 : 1 mixtures of **8a** and [H **39**]$^+$PF$_6^-$ and of [H **8a**]$^+$PF$_6^-$ and **39** in CD$_3$CN, whereas no free **8a** was observed in the similar experiments using proton sponge **40**. These results clearly indicated that azacalix[3]pyridine **8a** was more basic than proton sponge **40**. On the basis of the titration experiments using guanidine **39**, the pK_{BH+} value of azacalix[3]pyridine **8a** was determined to be 23.3 ± 0.1, implying that the basicity of **8a** was greater by a factor of 10^8 than

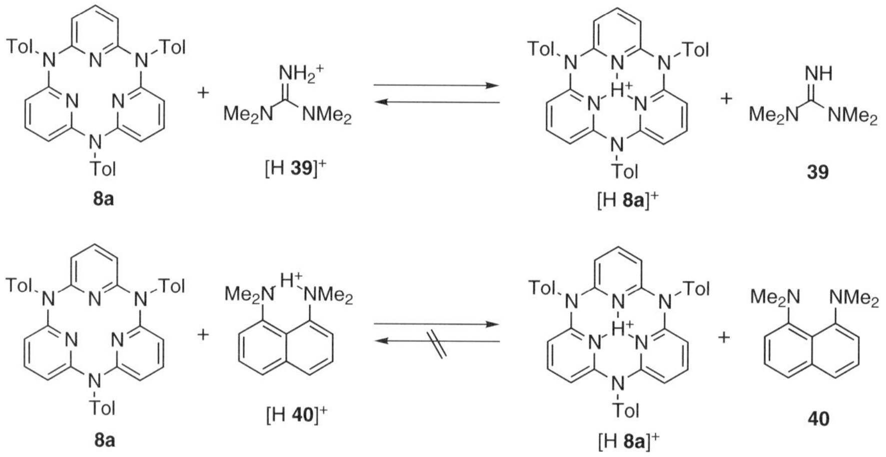

Scheme 12 Proton transportation experiments of azacalix[3]pyridine **8a** [37]

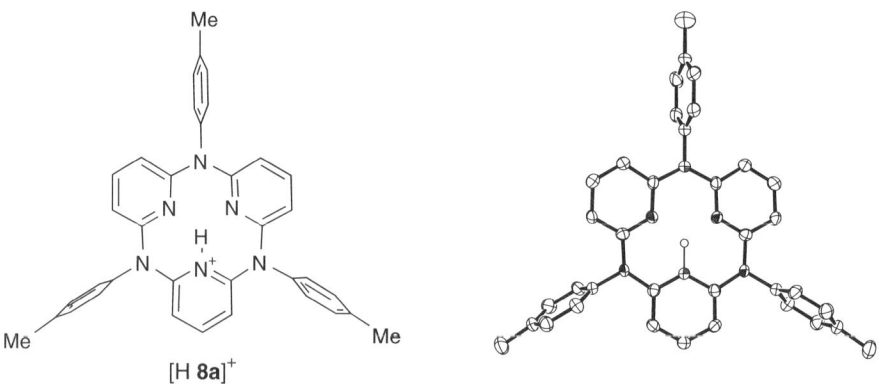

Fig. 13 X-ray crystal structures of monoprotonated azacalix[3]pyridine [H **8a**]$^+$ [37]

those of the component monomers such as 2-aminopyridine (pK_{BH+} = 14.26–14.66) and 2,6-diaminopyridine (pK_{BH+} = 14.56). This strong synergistic effect on protonation was demonstrated by the X-ray crystallographic analysis of the protonated species [H 8a]$^+$PF$_6^-$, in which one proton was embedded in the cavity and chelated by the nitrogen lone pairs of the pyridine rings, as shown in Fig. 13.

Gong et al. also investigated the protonation abilities of azacalix[n]pyridines 7b and 7f as well as azacalix[m]arene[m]pyridines 20b and 20f [24]. For studying protonation behavior, UV-vis titration experiments were carried out in order to estimate the protonation constants ($\log K_i$, where $i = 1$ to 8), as summarized in Table 2. Depending on the number of pyridine rings present, 7b, 7f, 20b, and 20f captured up to four, eight, two, and four protons, respectively. The more the pyridine rings present within a macrocycle, the larger the $\log K_i$ value. The bridging nitrogens exhibit low basicity, and no protonations were observed on them because of the conjugation and the steric hindrance. Exclusive protonations onto the pyridine nitrogens were supported by the X-ray crystallographic analyses of the protonated species of azacalix[4]pyridine 7b (Fig. 14).

Also interesting are the fullerene-complexation properties of azacalix[n]pyridines 7b, 7c, 7f, and 7g as well as azacalix[m]arene[m]pyridines 20b and 20f [24–26]. Depending on the size of the macrocyclic ring, azacalix[5]pyridine 7c, azacalix[8]pyridines 7f, azacalix[10]pyridine 7g, and azacalix[4]arene[4]pyridine 20f strongly interacted with fullerenes C_{60} and C_{70} in toluene, whereas smaller analogues 7b and 20b had no affinity towards them. The interaction between 7f or 20f and C_{60} was detectable even with the naked eye; the color of a toluene solution of C_{60} changed from its

Table 2 Protonation constants ($\log K_i$) of azacalix[n]pyridines and azacalix[m]arene[m]pyridines[a]

i	Protonation	$\log K_i$			
		7b ($n = 4$)	7f ($n = 8$)	20b ($m = 2$)	20f ($m = 4$)
1	L + H$^+$ ⇌ [HL]$^+$	8.4 ± 0.2	9.9 ± 0.3	5.84 ± 0.08	7.1 ± 0.7
2	[HL]$^+$ + H$^+$ ⇌ [H$_2$L]$^{2+}$	5.5 ± 0.7	7.6 ± 0.9	1.3 ± 0.0	4.9 ± 0.0
3	[H$_2$L]$^+$ + H$^+$ ⇌ [H$_3$L]$^{3+}$	3.4 ± 0.2	6.1 ± 0.1	–[b]	2.8 ± 0.9
4	[H$_3$L]$^+$ + H$^+$ ⇌ [H$_4$L]$^{4+}$	0.9 ± 0.2	5.0 ± 0.7	–[b]	1.0 ± 0.1
5	[H$_4$L]$^+$ + H$^+$ ⇌ [H$_5$L]$^{5+}$	–[b]	3.7 ± 0.3	–[b]	–[b]
6	[H$_5$L]$^+$ + H$^+$ ⇌ [H$_6$L]$^{6+}$	–[b]	2.2 ± 0.3	–[b]	–[b]
7	[H$_6$L]$^+$ + H$^+$ ⇌ [H$_7$L]$^{7+}$	–[b]	1.6 ± 0.4	–[b]	–[b]
8	[H$_7$L]$^+$ + H$^+$ ⇌ [H$_8$L]$^{8+}$	–[b]	1.3 ± 0.6	–[b]	–[b]

[a] Quoted from [24]
[b] No protonation

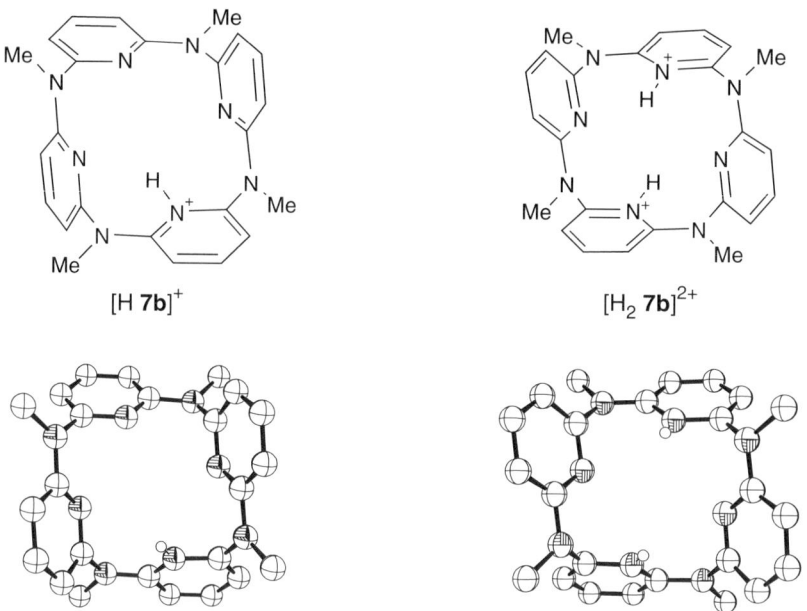

Fig. 14 X-ray crystal structures of mono- and diprotonated species of azacalix[4]pyridine 7b [24]

characteristic purple to a light brown. Fluorescence titration experiments were performed in order to estimate the stability constants (Table 3). As a result, azacalix[n]pyridines 7c, 7f, and 7g as well as azacalix[4]arene[4]pyridine 20f exhibited much higher stability constants than those obtained for the complexations of fullerenes with other mono-macrocyclic receptors thus re-

Table 3 Stability constants (10^5 mol^{-1} dm^3) for the complexation of fullerenes with azacalix[n]pyridines and azacalix[m]arene[m]pyridines[a]

Compound	Fullerene	
	C_{60}	C_{70}
7b ($n = 4$)	_[b]	_[b]
20b ($m = 2$)	_[b]	_[b]
7c ($n = 5$)	2.6 ± 0.01	1.2 ± 0.03
7f ($n = 8$)	4.6 ± 0.02	1.1 ± 0.02
20f ($m = 4$)	7.1 ± 0.02	13.7 ± 0.04
7g ($n = 10$)	3.0 ± 0.008	1.3 ± 0.03

[a] Quoted from [24–26]
[b] No complexation

ported. The efficient complexations of fullerenes by these macrocycles were interpreted in terms of the complementarity between the host and guest.

5
Concluding Remarks

A variety of azacalixarenes have been produced by applying three different synthetic strategies that are essentially the same as those established in calixarene chemistry. The diversity is still limited as compared with carbon- and sulfur-bridged calixarenes, however, structural perturbations imparted by the bridging nitrogen atoms have been clearly disclosed, especially through investigations regarding the solid state structures. The reason for this is obvious; the bridging nitrogen atoms are allowed to effectively conjugate with the aromatic π-systems due to the allocation of the nitrogen lone pair to the $2p$ orbital with a substantially identical shape to that of carbon atom. This electronic effect is also reflected to the complexation properties of azacalixarenes, in particular, with electron-deficient fullerenes. However, in order for azacalixarene chemistry to progress, a broader knowledge of this new calixarene family has to be further accumulated and organized, as the history of calixarenes and thiacalixarenes has demonstrated.

Acknowledgements We are grateful to the volume editor, Prof. Matsumoto K (Chiba Institute of Science), for his valuable comments on our manuscript. Our work in this review was partially supported by a Grant-in-Aid for Scientific Research (No. 19550037) from the Ministry of Education, Sports and Culture of Japan.

References

1. Gutsche CD (1989) In: Stoddart JF (ed) Calixarenes. Royal Society of Chemistry, Cambridge
2. Vicens J, Böhmer V (eds) (1991) Calixarenes: a versatile class of macrocyclic compounds. Kluwer, Dordrecht
3. Gutsche CD (1998) In: Stoddart JF (ed) Calixarenes revisited. Royal Society of Chemistry, Cambridge
4. Asfari Z, Böhmer V, Harrowfield J, Vicens J, Saadioui M (eds) (2001) Calixarenes 2001. Kluwer, Dordrecht
5. König B, Fonseca MH (2000) Eur J Inorg Chem 2303
6. Lhoták P (2004) Eur J Org Chem 1675
7. Morohashi N, Narumi F, Iki N, Hattori T, Miyano S (2006) Chem Rev 106:5291
8. Smith GW (1963) Nature 198:879
9. Gutsche CD, Iqbal M (1979) Org Synth 68:234
10. Gutsche CD, Dhawan B, Leonis M, Stewart D (1979) Org Synth 68:238
11. Munch JH, Gutsche CD (1979) Org Synth 68:243
12. Wolfe JP, Wagaw S, Marcoux JF, Buchwald SL (1998) Acc Chem Res 31:805

13. Hartwig JF (1998) Acc Chem Res 31:852
14. Hartwig JF (1998) Angew Chem Int Ed 37:2046
15. Muci AR, Buchwald SL (2002) Top Curr Chem 219:133
16. Ito A, Ono Y, Tanaka K (1998) New J Chem 779
17. Ito A, Ono Y, Tanaka K (1999) J Org Chem 64:8236
18. Miyazaki Y, Kanbara T, Yamamoto T (2002) Tetrahedron Lett 43:7945
19. Suzuki Y, Yanagi T, Kanbara T, Yamamoto T (2005) Synlett 263
20. Hayes BT, Hunter RF (1956) Chem Ind 193
21. Hayes BT, Hunter RF (1958) J Appl Chem 8:743
22. Tsue H, Ishibashi K, Takahashi H, Tamura R (2005) Org Lett 7:2165
23. Böhmer V, Chhim P, Kämmerer H (1979) Makromol Chem 180:2503
24. Gong HY, Zhang XH, Wang DX, Ma HW, Zheng QY, Wang MX (2006) Chem Eur J 12:9262
25. Liu SQ, Wang DX, Zheng QY, Wang MX (2007) Chem Commun 3856
26. Wang MX, Zhang XH, Zheng QY (2004) Angew Chem Int Ed Engl 43:838
27. Wang MX, Yang HB (2004) J Am Chem Soc 126:15412
28. Graubaum H, Lutze G, Costisella BJ (1997) J Prakt Chem/Chem-Ztg 339:266
29. Graubaum H, Lutze G, Costisella BJ (1997) J Prakt Chem/Chem-Ztg 339:672
30. Fukushima W, Kanbara T, Yamamoto T (2005) Synlett 2931
31. Ishibashi K, Tsue H, Tokita S, Matsui K, Takahashi H, Tamura R (2006) Org Lett 8:5991
32. Hartwig JF, Richards S, Barañano D, Paul FJ (1996) J Am Chem Soc 118:3626
33. Tsue H, Ishibashi K, Tokita S, Matsui K, Takahashi H, Tamura R (2007) Chem Lett 36:1374
34. Sone T, Ohba Y, Moriya K, Kumada H, Ito K (1997) Tetrahedron 53:10689
35. Araki K, Iwamoto K, Shinkai S, Matsuda T (1989) Chem Lett 1747
36. Iwamoto K, Araki K, Shinkai S (1991) J Org Chem 56:4955
37. Kanbara T, Suzuki Y, Yamamoto T (2006) Eur J Org Chem 3314

Chemistry of Calixfurans

Kei Goto

Interactive Research Center of Science, Graduate School of Science and Engineering, Tokyo Institute of Technology, 2-12-1 Ookayama, Meguro-ku, 152-8551 Tokyo, Japan
goto@chem.titech.ac.jp

Dedicated to Professor Renji Okazaki on the occasion of his 70th birthday.

1	Introduction	98
2	Syntheses of Calix[n]furans	99
2.1	Syntheses of Calix[4]furans	99
2.1.1	Single-Step Synthesis	99
2.1.2	Stepwise Synthesis	101
2.2	Syntheses of Calix[5]furans, Calix[6]furans, and Larger Homologues	105
2.2.1	Single-Step Synthesis	105
2.2.2	Stepwise Synthesis	105
3	Reactions of Calix[n]furans	108
3.1	Transformations to Other Heterocalixarenes	108
3.2	Modification via Cycloaddition Reactions	112
4	Structures and Inclusion Properties of Calixfurans	113
4.1	Structures of Calixfurans	113
4.2	Inclusion Properties of Calixfurans and Their Derivatives	117
5	Concluding Remarks	119
References		119

Abstract The extensive development of the studies on calixarenes has promoted a growing interest in exploring the chemistry of heterocalixarenes in which the phenolic units of calixarenes are replaced by heterocycles. Calixfurans, or tetraoxaporphyrinogens, constitute a major class of heterocalixarenes. In addition to their potential capability as receptor molecules, they have been employed as a versatile molecular platform for further chemical transformation into a variety of macrocyclic compounds by taking advantage of the chemical lability of the furan units. This review summarizes the synthesis, reactions, structures, and host–guest chemistry of calix[n]furans and their hybrid systems containing other aromatic units such as pyrrole and thiophene.

Keywords Calixfurans · Conformation · Crystal structure · Heterocalixarenes · Inclusion Properties

1
Introduction

Calixarenes are cyclic compounds made up of benzene rings connected by methylene bridges in *meta*-positions, which were named by Gutsche from the Greek calix meaning "vase" or "chalice" [1–5]. They have been widely utilized as versatile building blocks in supramolecular chemistry. The extensive development of studies on calixarenes has promoted a growing interest in exploring the chemistry of analogues of this class of macrocycles, in which the phenolic units are replaced by heterocycles [6–9]. Among such heterocalixarenes, calixpyrroles have attracted a lot of attention because of their applications in designing receptors and molecular devices, their oxidation to porphyrins, and the fact that many review articles have been published about their chemistry [10–12]. Calixfurans, or tetraoxaporphyrinogens, also constitute a major class of heterocalixarenes, the history of which dates back to the beginning of 1900s. The cyclic array of furan rings forms a relatively π-electron-rich cavity. The structures of calixfuran macrocycles somewhat resemble those of crown ethers, although the donor ability of furan oxygen is weaker than that of ether oxygen. In addition to the potential capability of calixfurans as receptor molecules, the furan rings in their framework have high chemical lability, which enables them to be used as a building block in organic synthesis [13, 14]. Containing such reactive rings as the main components, calixfurans can serve as a versatile molecular platform for further chemical transformation into a variety of macrocyclic compounds. In the present review, the synthesis, reactions, structures, and host–guest chemistry of calix[n]furans are discussed. For simplicity, the cyclic oligomers of furan units, i.e., calix[n]furans, and the linear oligomers are denoted as Cn and Ln in some parts of this article.

2
Syntheses of Calix[*n*]furans

2.1
Syntheses of Calix[4]furans

2.1.1
Single-Step Synthesis

The preparation of calixfurans by one-pot condensation of furan with ketones has been investigated since the 1950s. It was revealed, however, that the first single-step synthesis of a calixfuran was unintentionally carried out in the beginning of 1900s, when Hale et al. treated ethyl furoate with ethyl magnesium iodide [15]. The product was originally characterized as 3-furyl-2-pentane by Hale et al. but later identified to be calix[4]furan **1** by Ackman et al. [16], as well as Beals et al. [17].

1: R=R'=Me
2: R,R'= ⌬
3: R=Me, R'=Et
4: R=R'=H

In 1955, Ackman et al. reported the preparation of calix[4]furans by acid-promoted condensation of furan with ketones [16]. When acetone and furan were condensed in the presence of hydrochloric acid, calix[4]furan **1** was obtained in 18% yield. Chastrette et al. found that the addition of metal salts increased the yield of **1**, e.g., 25% in the presence of $LiClO_4$, suggesting that a template effect may be operating [18]. However, by close examination of the reaction conditions, Rest et al. proposed that the action of metal salts can be explained in terms of their effect on the acidity of the reaction medium rather than the ion template effect [19, 20]. They also found that the yields of calix[4]furans are dependent on the concentration of hydrochloric acid used for condensation. The maximum yields of 35% and 16% were obtained for the acetone-condensed product **1** and the cyclohexanone-condensed product **2**, respectively. Jurczak et al. reported that the use of highly concentrated sulfuric acid afforded the calix[4]furanes **1** and **2** in yields of 71% and 49%, respectively, although no metal salt was added [21]. Also in these cases, a substantial effect of acid concentration on the reaction yield was observed. The highest yields were obtained for concentration of sulfuric acid 90.5% and 87.6% for **1** and **2**, respectively. However, cyclocondensation reactions under similar conditions using other ketones (such as chloroacetone, various methyl alkyl ketones, acetophenone, as well as higher symmetric ketones) were unsuccessful. For example, successful one-pot synthesis of **3** from furan and 2-butanone has never been described in the literature.

While the first synthesis of the parent calix[4]furan **4** was carried out by the stepwise procedure [22], Vogel et al. reported the single-step synthesis of **4** by cyclocondensation of furfuryl alcohol (**5**), not by the reaction of furan with formaldehyde or its derivatives (Scheme 1) [23]. They reported that treatment of **5** with $ZnCl_2/HCl$ produced **4**, which was isolated only in 1% yield but was made tolerable for a preparative scale reaction through an easy purification procedure. Cyclocondensation of **5** catalyzed by $BF_3 \cdot Et_2O$ also afforded **4** in 1.3% yield. The high yield synthesis of **4** was achieved by the stepwise procedure using $BF_3 \cdot Et_2O$ (vide infra) [24].

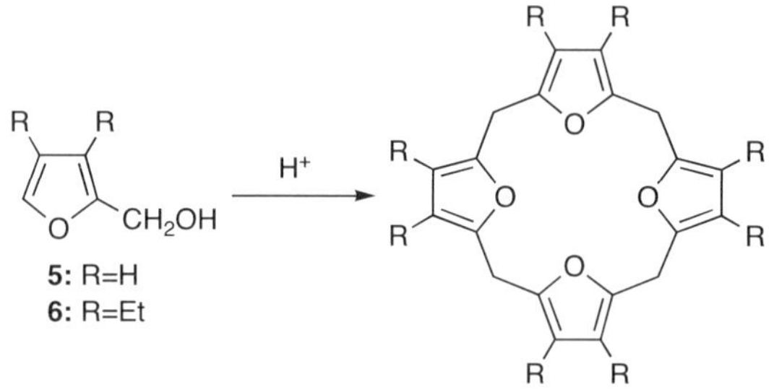

Scheme 1 Syntheses of calix[4]furans **4** and **7** [23, 24]

When furfuryl alcohol **6** bearing ethyl groups on the furan ring was subject to cyclocondensation reaction in the presence of p-toluenesulfonic acid, the corresponding calix[4]furan **7** was obtained in 16% yield (Scheme 1) [25], which is much better than that of the conversion of **5** to the parent macrocycle **4**. This yield enhancement was explained in terms of the "helical effect", in which the ethyl groups promote ring closure to form the macrocycle on conformational grounds. In addition to **7**, the larger cyclocondensation products with five or more furan components are formed (total 2–3%).

Another unique one-pot synthesis of calixfuran analogues is that of the silicon-bridged calixfurans. Treatment of furan with two equimolar amounts of n-BuLi/TMEDA/t-BuOK followed by slow addition of Me_2SiCl_2 afforded the cyclic tetramer **8** and hexamer **9** in 16% and 10% yields, respectively (Scheme 2) [26, 27].

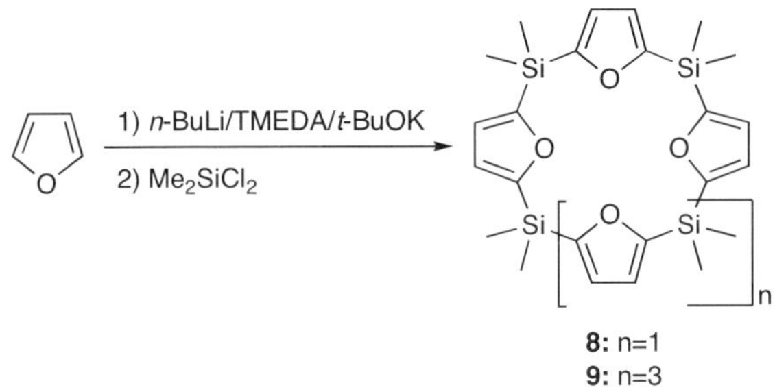

Scheme 2 Syntheses of the silicon-bridged calixfurans 8 and 9 [26, 27]

2.1.2
Stepwise Synthesis

In the article describing the single-step synthesis of calix[4]furans, Ackman et al. also reported the HCl-promoted cyclocondensation of the linear tetramer 10 bearing isopropylidene bridges with acetone, 2-butanone, and 3-pentanone to produce calix[4]furans 1, 13, and 14 in 47–53% yields (Scheme 3) [16]. The cyclocondensation of the L4 unit 10 with other carbonyl compounds such as chloroacetone, levulinic acid, ethyl levulinate, and cyclohexanone afforded the corresponding non-symmetric calix[4]furans such as 15 (Scheme 3) [28, 29]. The isobutylidene-bridged calix[4]furan 3, which was not accessible by the single-step procedure, was also obtained by the L4 unit 11 bearing isobutylidene bridges with 2-butanone in 65% yield (Scheme 3) [30]. The same product was obtained in 37% yield at the highest by the fragment coupling method with the corresponding L2 unit 17 and 2-butanone [16].

The parent calix[4]furan 4 was first isolated by Vogel et al. in the reaction of the L2 fragment 18 with formaldehyde in the presence of $LiClO_4$ and perchloric acid, albeit in low yield of 0.5–1% [22]. Preparation of 4 in reasonable yield was attained by the Lewis acid-catalyzed cyclization of the L4 precursor 12. Musau and Whiting reported that the linear tetramer 12 was converted into 4 by using $CH_2(OMe)_2$ instead of formaldehyde as the bridging reagent in the presence of $BF_3 \cdot OEt_2$ in dichloromethane [24]. The analysis of the crude product showed that the reaction was essentially quantitative. However, the chromatographic purification caused considerable product decomposition to reduce the isolated yield of 4 to 34%. In the reaction of the L2 fragment 18 with $CH_2(OMe)_2$ under similar conditions, calix[4]furan 4 was obtained only in 6% yield. They noted that the use of HCl as a condensation catalyst resulted in the decomposition of the CH_2-bridged linear oligomers.

Although the cyclization of the L4 precursor is promising in general, the coupling of L2 fragments also provides an efficient route to calix[4]furans because a variety of L2 units are easily accessible in comparison with longer

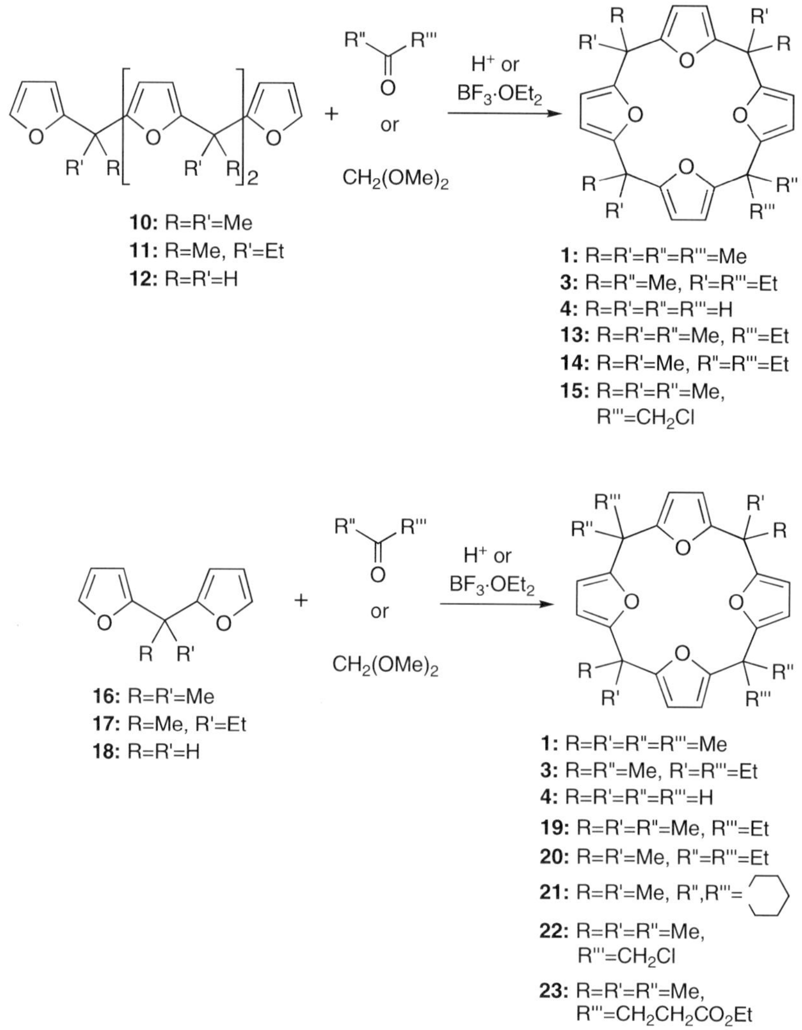

Scheme 3 Syntheses of calix[4]furans by the stepwise methods [16, 22, 24, 28–30]

oligomers. This approach was first described by Ackman et al. again, who reported the HCl-catalyzed condensation of the L2 units **16** or **17** bearing an isopropylidene or isobutylidene bridge with acetone or 2-butanone [16]. The corresponding calix[4]furans **1**, **3**, and **19** were obtained in 37–43% yields. By cyclocondensation of L2 units with ketones under similar conditions, Brown et al. synthesized calix[4]furans with various bridging unit such as cyclohexylidene bridges and chloromethyl-substituted bridges [17, 28, 30, 31]. Jurczak et al. also used highly concentrated sulfuric acid for the cyclocondensation of L2 fragments with ketones [29]. Calix[4]furans **19-23** with various substituents as well as the octamethyl derivative **1** were obtained efficiently in 11–66% yields, which are generally higher than those of the corresponding HCl-catalyzed reactions.

There was no report on the synthesis of calixfurans by cyclocondensation of furan with aldehyde in the single-step procedure. However, the condensation of the L2 unit **24** with acetaldehyde in the presence of $LiClO_4/HClO_4$ afforded the calix[4]furan **25** bearing ethylidene bridges in 31% yield as a mixture of the stereoisomers (Scheme 4) [32, 33]. Calix[4]furans bearing propylidene or benzylidene bridges were also prepared by similar methods [33, 34].

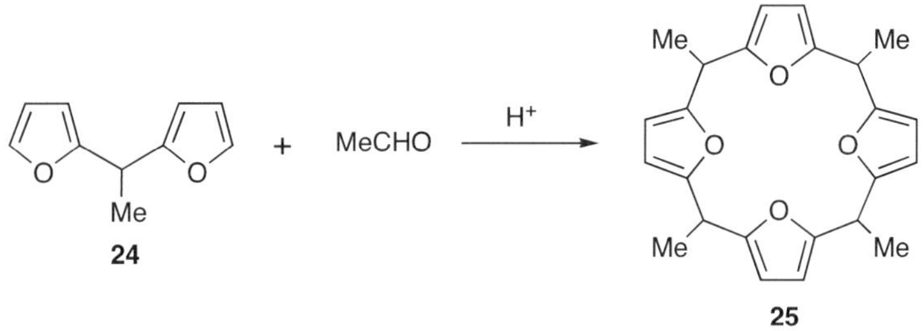

Scheme 4 Synthesis of calix[4]furan **25** by the condensation of the L2 unit with aldehyde [32, 33]

Other types of L2 fragment coupling involves the reaction of L2 units bearing hydroxymethyl or alkenyl groups at both ends. Thus the reaction of diol **26** and difurylalkane **16** in the presence of hydrochloric acid afforded calix[4]furan **1** although in low yield (Scheme 5) [31]. Calix[4]furan **2** was obtained by the coupling of the L2 unit **27** bearing cyclohexenyl groups with difurylalkane **28** [28].

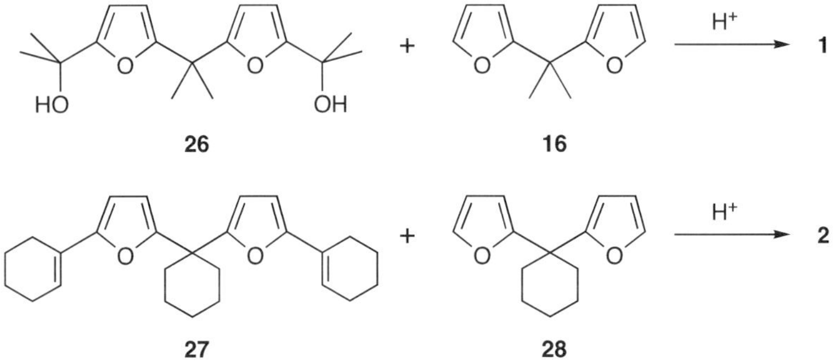

Scheme 5 Syntheses of calix[4]furans by the L2 fragment coupling method [28, 31]

Such methodology has been conveniently employed in the synthesis of hybrid calixfuran systems containing other aromatic units such as pyrrole

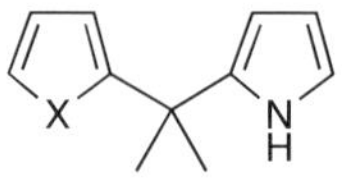

29: X=O
30: X=NH

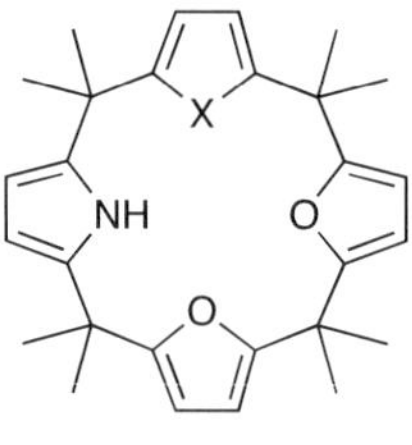

31: X=O
32: X=NH

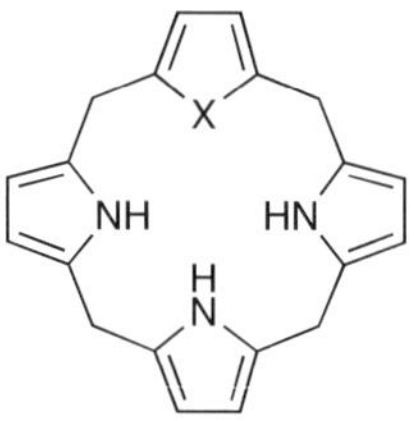

33: X=O
34: X=NH

and thiophene. In the chemistry of heterocalixarenes, incorporation of various kinds of heterocycles as a part of the parentmacrocycle has been an effective means to modulate the binding properties of the receptors. A variety of hybrid calixfurans have been synthesized by the coupling of linear oligomers with their counterparts bearing α-hydroxyalkyl groups at both ends. In 1958, Brown et al. reported the first synthesis of hybrid calix[4]furans by the reaction of diol **26** with furylpyrrolylmethane **29** or dipyrrolylmethane **30** to produce calix[3]furan[1]pyrrole **31** or calix[2]furan[2]pyrrole **32**, respectively [35]. Lee et al. synthesized various kinds of hybrid calix[4]furans containing pyrrole and/or thiophene rings by BF_3-catalyzed (2 + 2) condensation of the L2 unit with the L2-derived diol [36] or (3 + 1) condensation of the L3 unit with the L1-derived diol [37, 38]. For example, calix[1]furan[2]pyrrole[1]thiophene **37** was prepared by the BF_3-catalyzed condensation of the hybrid L3 unit **35** with diol **36** in 39% yield (Scheme 6).

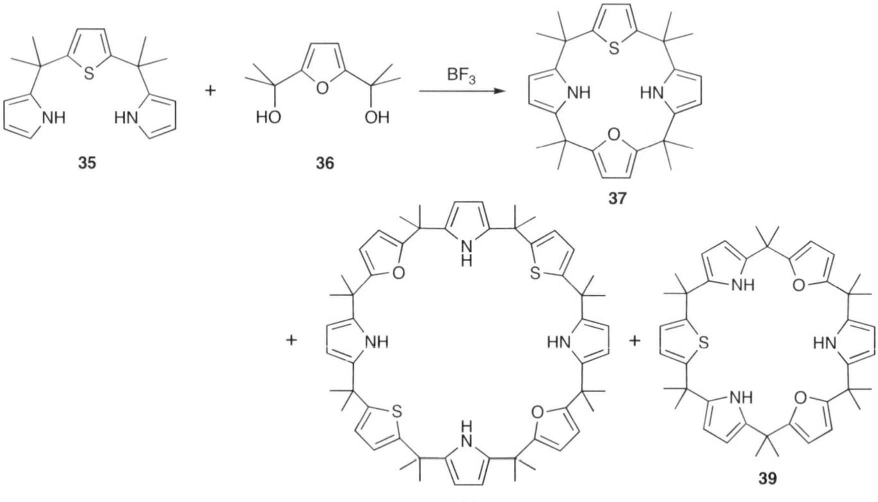

Scheme 6 Syntheses of hybrid calixfurans by the L2 fragment coupling method [38]

In this reaction, the corresponding cylic octamer **38** (11%) as well as hexamer **39** (2%) was also obtained. The formation of hexamer **39** was explained in terms of acid-catalyzed, reversible cleavage of the starting material during the reaction. The unsubstituted calix[1]furan[3]pyrrole **33** and calix[4]pyrrole **34** were isolated for the first time by Taniguchi et al in the similar BF_3 catalyzed (3 + 1) condensation [39].

2.2
Syntheses of Calix[5]furans, Calix[6]furans, and Larger Homologues

2.2.1
Single-Step Synthesis

In the acid-catalyzed condensation of furan and acetone, not only calix[4]-furan **1** but also the larger homologues are generated. With the furan to acetone ratio of 1 : 6, the reaction products included C4 **1**, C5 **40**, and C6 **41** in the ratio of ca. 12.5 : 1 : 1.2 [40]. An analysis of different crude mixtures obtained by varying the ratio of furan to acetone indicated that an excess of acetone favors the formation of C4 **1** over linear oligomers, and it also revealed that C4 **1** is always the major cyclic component with respect to the larger macrocycles **40** and **41**.

40: n=1, R=Me
41: n=2, R=Me
42: n=4, R=Me
43: n=5, R=Me
44: n=1, R=H
45: n=2, R=H
46: n=3, R=H
47: n=4, R=H

48: n=1, R=Me
10: n=2, R=Me
49: n=3, R=Me
50: n=4, R=Me
51: n=7, R=Me
52: n=1, R=H
12: n=2, R=H
53: n=3, R=H
54: n=4, R=H

2.2.2
Stepwise Synthesis

Calix[5]furan **40** with isopropylidene bridges was prepared by the HCl-catalyzed cyclocondensation of the L5 unit **49** with acetone in 45% yield [32, 41]. An approach to **40** by the coupling of the L2 and L3 units was not very successful. In the condensation of the L2 unit **16**, the L3 unit **48**, and acetone with and without $LiClO_4 \cdot (DME)_2$, the ratios of C4 **1** : C5 **40** : C6 **41** were 2.8 : 1 : 1.8 and 4.2 : 1 : 1.4, respectively [40]; C5 **40** is always the least favored product. This result also indicates that the efficient synthesis of the larger calixfurans directly from furan and acetone is hampered by the fact that C4 **1** constitutes a sink for the growing oligomeric chain. The parent calix[5]furan

44 was obtained by cyclocondensation of the L5 unit **53** with $CH_2(OMe)_2$ in the presence of $BF_3 \cdot OEt_2$ in 3–5% yield [24].

Calix[6]furan **41** with isopropylidene bridges was also obtained by the cyclocondensation of the linear precursor **50** with acetone in 52% yield [32]. Addition of $LiClO_4$ or $CsClO_4$ showed essentially no effect on the yield in the hydrochloric acid-promoted synthesis of **41** [20].

Preparation of C6 **41** by the condensation of the L3 unit **48** and acetone was first reported by Ackmann et al., although in low yield [16]. Kohnke et al. improved the procedure and isolated C6 **41** in 25–28% yield [40, 42]. While the yield is not so high, their protocol does not require chromatographic purification, which is suitable for a large-scale preparation. Jurczak et al. reported the condensation of L3 **48** with cyclohexanone or ethyl levulinate in the presence of 90.5% sulfuric acid to produce the corresponding non-symmetric calix[6]furans **55** and **56** (Scheme 7) [29]. The parent calix[6]furan **45** was obtained by cyclocondensation of the L6 unit **54** with $CH_2(OMe)_2$ in the presence of $BF_3 \cdot OEt_2$, although only in 1% [24].

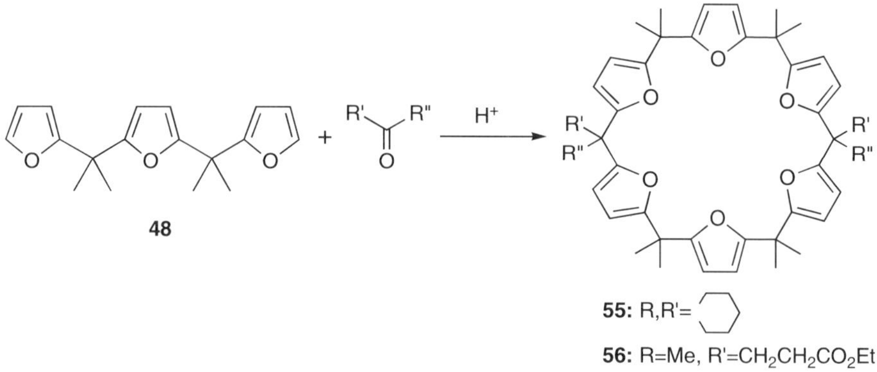

55: R,R'= ⬡

56: R=Me, R'=$CH_2CH_2CO_2Et$

Scheme 7 Syntheses of calix[6]furans **55** and **56** by the condensation of the L3 unit with ketone [29]

Isolation of calix[7]furan derivatives has never been reported to date although the parent calix[7]furan **46** was detected in the crude mixture of the reaction of the L2 unit **18**, the L3 unit **52**, and $CH_2(OMe)_2$ in the presence of $BF_3 \cdot OEt_2$ [24].

Calix[8]furan derivatives have only been obtained as minor products in the condensation of L4 units [20]. In the HCl-promoted condensation of the L4 unit **10** and acetone, which afforded C4 **1** as the major product, C8 **42** was also obtained, albeit only in low yield. Similarly, the parent calix[8]furan **47** was obtained in 2% yield in addition to C4 **4** (34%) in the condensation of the L4 unit **12** with $CH_2(OMe)_2$ in the presence of $BF_3 \cdot OEt_2$ [24].

Calix[9]furan **43** with isopropylidene bridges was prepared by the HCl-promoted cyclocondensation of the L9 precursor **51** with acetone in 45%

yield [40]. Under the same conditions, the condensation of the L3 unit **48** and acetone afforded C6 **41** and C9 **43** in 18% and 6.5% yields, respectively.

For the preparation of larger family of calixfurans, the cyclization of the corresponding linear oligomers is considered to be the most promising way when such precursors are available. Although not reported yet, the cyclization of the L7 and L8 units with isopropylidene bridges, which were prepared

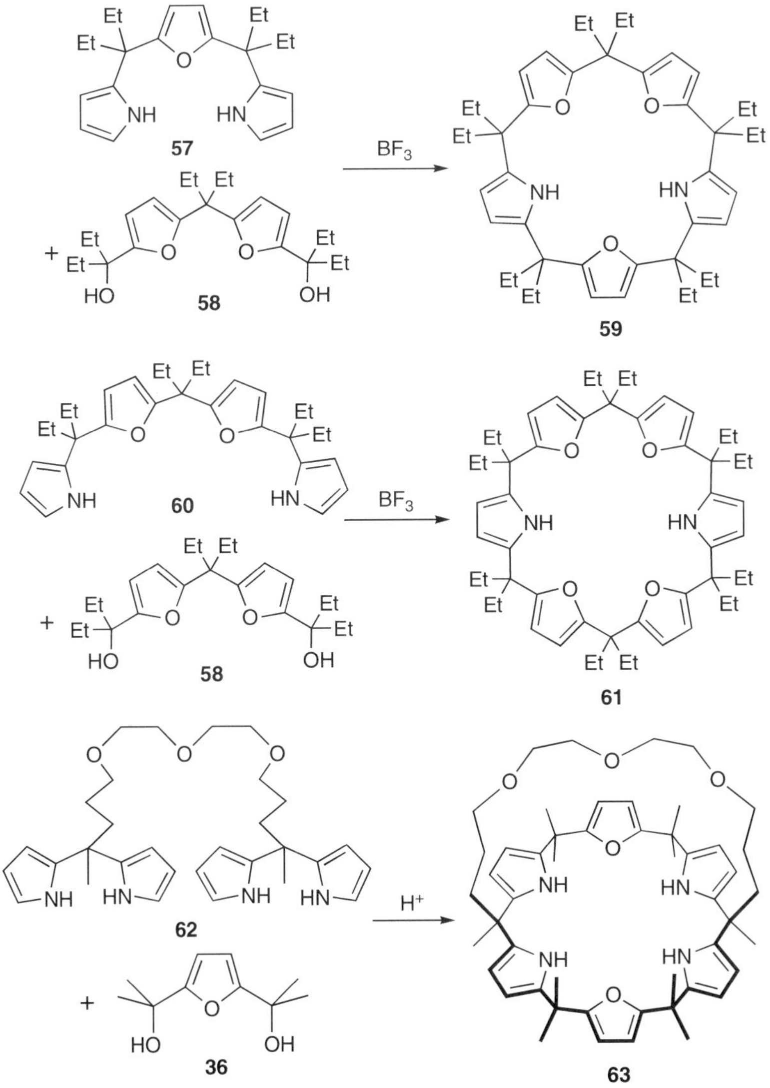

Scheme 8 Syntheses of hybrid calixfurans by the coupling of linear oligomers [43, 44]

by Rees et al. by a rational approach [41], would afford the corresponding calix[7]furan and calix[8]furan, respectively, in reasonable yields.

Similar to hybrid calix[4]arenes, the larger family of hybrid calixfurans containing pyrrole and/or thiophene rings in various patterns have been prepared by the condensation of linear oligomers with their counterparts bearing α-hydroxyalkyl groups. Lee et al. reported the BF_3 catalyzed (3 + 2) condensation of the L3 unit **57** with the L2-derived diol **58** to produce calix[3]furan[2]pyrrole **59** (55%), in which the corresponding calix[6]furan[4]pyrrole (15%) was also obtained by (3 + 2 + 3 + 2) condensation (Scheme 8) [43]. In the (4 + 2) condensation of the L4 unit **60** and diol **58**, calix[4]furan[2]pyrrole **61** and the corresponding calix[8]furan[4]pyrrole were obtained in 53% and 11% yield, respectively. The cryptand-like calix[2]furan[4]pyrrole **63** having a polyether strap was synthesized by the condensation of bis(bipyrrylmethane) **62** with diol **36** in the presence of trifluoroacetic acid in 22% yield [44].

3
Reactions of Calix[n]furans

3.1
Transformations to Other Heterocalixarenes

One of the synthetic utilities of furans is their ability to function as masked 1,4-dicarbonyl compounds. In 1981, Le Goff and Williams reported the oxidative ring-opening of the furan units of calix[4]furan **1** and calix[6]furan **41** [45]. Oxidation of calix[4]furan **1** by bromine in aqueous acetic acid resulted in the furan-ring opening to produce bis(trans-enedione) **64** (Scheme 9). In this reaction, the use of 4 equimolar amount of bromine did not open more than two of the furan rings. However, oxidation of **1** by 4.2 equimolar amount of m-chlorobenzoic acid (MCPBA) afforded the tetra-ring-opened octaketone **65**, in which cis-enediones were formed stereospecifically. Likewise, treatment of calix[6]furan **41** with 6.3 equimolar amount of MCPBA produced dodecaketone **67**. Similarly, decaketone **66** was also obtained by MCPBA oxidation of calix[5]furan **40** [46].

By varing the stoichiometry of MCPBA in these reactions, partially ring-opened products can also be obtained. For example, calix[4]furan **1** reacted with 3.1 equimolar amount of MCPBA to afford the tri-ring-opened product **68**, while treatment with 2.2 equimolar amount of the peracid produced a mixture containing the di-ring-opened regioisomers **69** and **70**, as well as **68** [45]. Oxidation of calix[6]furan **41** with 4 equimolar amount of MCPBA gave a mixture of the enediones **71**, **72**, and **73** [42].

The endiones such as **64–73** obtained by bromine- or MCPBA-mediated ring opening of calixfurans can be reduced to the corresponding saturated

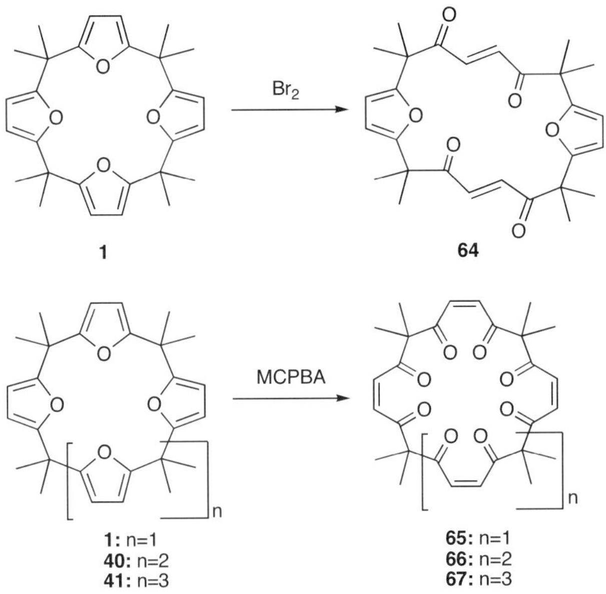

Scheme 9 Oxidation of calixfurans [45, 46]

derivatives, such as **74** and **75**, which accomplishes formal hydrolysis of the furan units in the macrocycle to 1,4-diketones. The 1,4-diketone derivatives thus obtained can be subjected to the Pall–Knorr pyrrole synthesis to produce a macrocycle containing pyrrole rings. Tetraketone **75** derived from calix[4]furan **1** via **64** was treated with ammonium acetate to give calix[2]furan[2]pyrrole **76** [47].

By homologation of the furan rings of calix[5]furan **40** and calix[6]furan **41** to pyrrole, Kohnke et al. reported the syntheses of the larger family of calixpyrroles, which otherwise are not readily obtainable, as well as the hybrid systems. In contrast with calix[4]pyrroles such as **77**, the larger family of calixpyrroles such as **78** and **79** tend to undergo a facile mitosis reaction to the cyclic tetramer even under mild acidic conditions. Reduction of the calix[5]furan-based enedione **66** followed by Pall–Knorr reaction afforded calix[5]pyrrole **78**, which was found not to be so unstable as originally envisaged (Scheme 10) [46]. Similarly, calix[1]furan[5]pyrrole **80**, calix[2]furan[4]pyrrole **81**, calix[3]furan[3]pyrrole **82**, and calix[6]pyrrole **79** were derived from the calix[6]furan-based enediones **71**, **72**, **73**, and **67**, respectively [42, 48].

Kohnke et al. also reported the conversion of calixfurans to heterocyclophanes containing isopyrazole units via cyclic polyketone derivatives such as **74** [49]. Treatment of polyketone **74** with hydrazine hydrate gave the isopyrazole-based macrocycles **83**. The corresponding hexamer was also prepared by the same protocol. The transformation of furans into thiophenes with hydrogen sulfide in acidic media is a well-established reaction in furan chemistry. By applying this reaction to the parent calix[4]furan **4**, Vogel et al. succeeded in the first synthesis of the parent calix[4]thiophene **84**, which has not been obtained by coup-

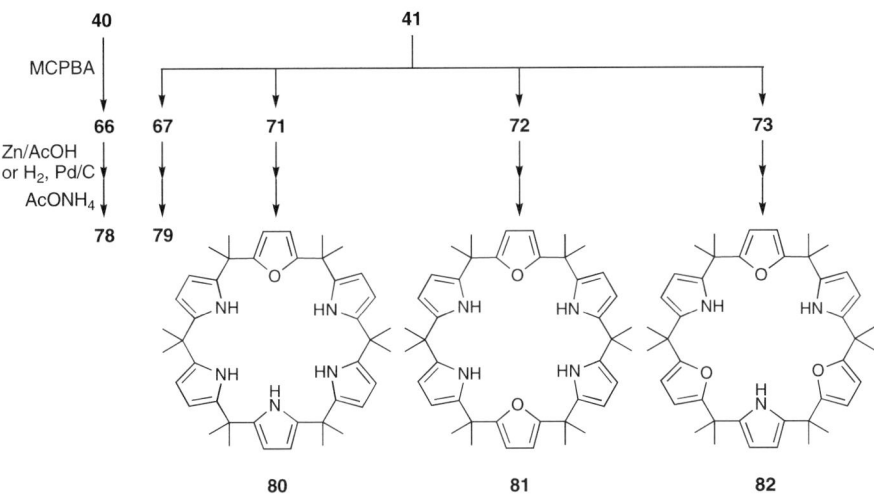

Scheme 10 Conversion of calixfurans to calixpyrroles and hybrid systems [42, 46, 48]

ling suitable 2,2′-dithienylmethane derivatives [23]. They also synthesized calix[4]selenophene **85** by the reaction of **4** with hydrogen selenide [23]. Ree et al. reported the conversion of calix[4]furan **1** and calix[6]furan **41** into macrocyclic isothiazoles, such as **86**, where the recently developed transformation of 2,5-disubstituted furans into 5-acyl-3-substituted isothiazoles was applied [50].

Oxidation of calix[4]furans **4** and **7** by nitric acid, cerium(IV) ammonium nitrate, or 2,3-dichloro-5,6-dicyano-1,4-benzoquinone (DDQ) followed by subsequent treatment with $HClO_4$ provided the corresponding dications **87** and **88**, respectively (Scheme 11) [23, 25, 51]. Catalytic hydrogenation of calixfurans on Raney nickel, Ru/C, or Ru–Rh/C produces the corresponding calixtetrahydrofurans, such as **89** [18, 32].

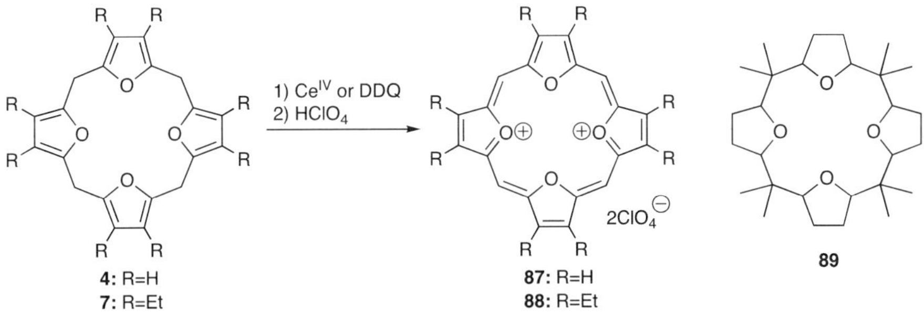

Scheme 11 Syntheses of the dication salts **87** and **88** [23, 25, 51]

3.2
Modification via Cycloaddition Reactions

The furan units of calixfurans are readily subjected to Diels–Alder reactions with various dienophiles. In 1982, Hart et al. reported the reaction of calix[4]furan **1** with 4 equimolar amount of benzyne (generated from benzenediazonium carboxylate hydrochloride) to give the adduct **90** (Scheme 12) [52]. Deoxygenation of **90** to produce the corresponding calix[4]naphthalene **91** was unsuccessful probably because of the rigidity or steric hindrance of **90**. On the other hand, such transformation from the furan ring to naphthalene was performed in the partially ring-opened calix[4]furan derivative **75** with a more flexible framework. Calix[2]furan[2]naphthalene **94** was prepared by benzyne addition to the bisfuran macrocycle **69** to produce adduct **92** followed by catalytic hydrogenation and subsequent acid-promoted dehydration [52].

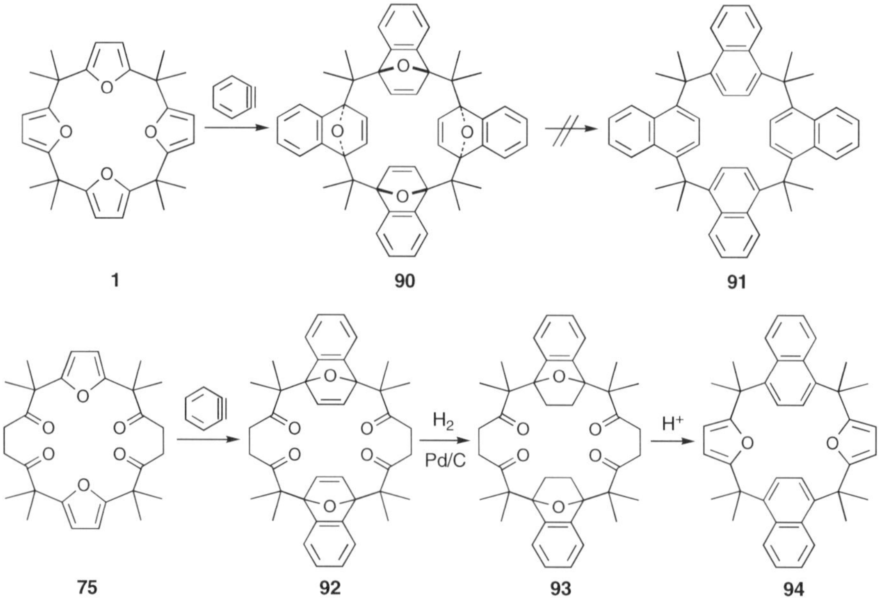

Scheme 12 Modification of calix[4]furan **1** and its derivative **75** via benzyne addition [52]

Chemical modification of calix[6]furan **41** by Diels–Alder reactions with benzyne and dimethyl acetylenedicarboxylate (DMAD) was studied by Kohnke et al. (Scheme 13) [53]. The reaction of **41** with benzyne afforded the adduct **95**, which was converted to calix[5]furan[1]naphthalene **96** by hydrogenation and subsequent dehydration. Two isomeric calix[4]furan[2]-naphthalenes **97** and **98** were obtained by similar chemical transformation starting from the corresponding bis-benzyne-adducts. Preparation of the tris-benzyne-adduct **99** was also reported.

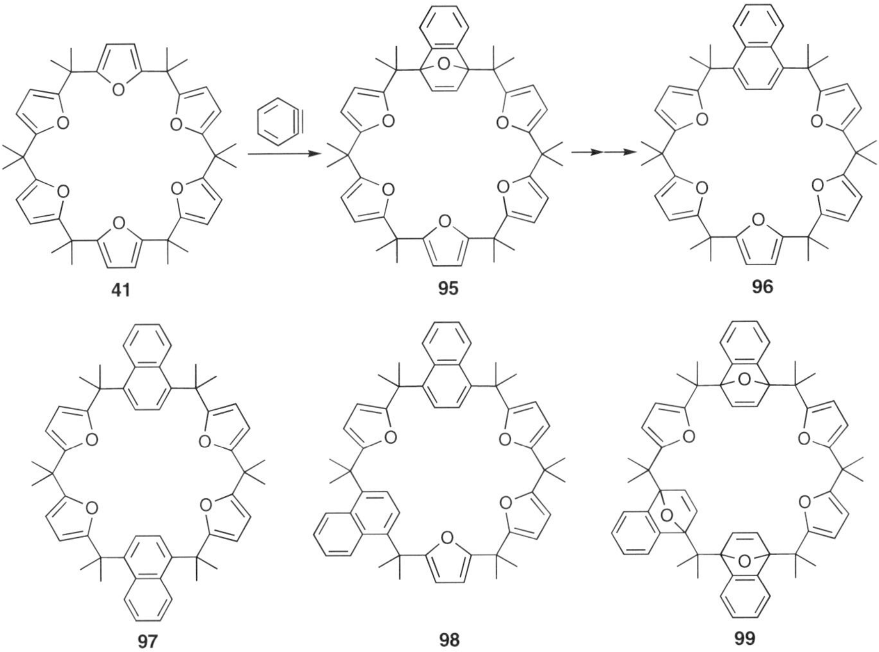

Chemistry of Calixfurans

Scheme 13 Modification of calix[6]furan **41** via benzyne addition [53]

The Diels–Alder reaction of calix[6]furan **41** with DMAD afforded the mono-adduct **100** and four isomeric bis-adducts **101** and **102** (syn and anti isomers for each) (Scheme 14) [53]. The oxanorbornadiene units of the adducts **100–102** were converted into 3,4-furandicarboxylate by the selective hydrogenation of the double bonds without the methoxycarbonyl groups followed by the retro Diels–Alder reaction. Thus, the calix[6]furan derivatives **103–105** were obtained via 3 steps starting from **41**. It is notable that calixfurans containing 3,4-furandicarboxylate units cannot be obtained by the condensation of 3,4-furandicarboxylate due to the electron-withdrawing deactivating effect of these groups.

4
Structures and Inclusion Properties of Calixfurans

4.1
Structures of Calixfurans

In general, there are four typical conformations for calix[4]arene derivatives, i.e., the cone, the partial cone, the 1,2-alternate, and the 1,3-alternate conformations (Fig. 1). It has been indicated that the parent calix[4]arene **106** predominantly takes the cone conformation both in the solid state and in so-

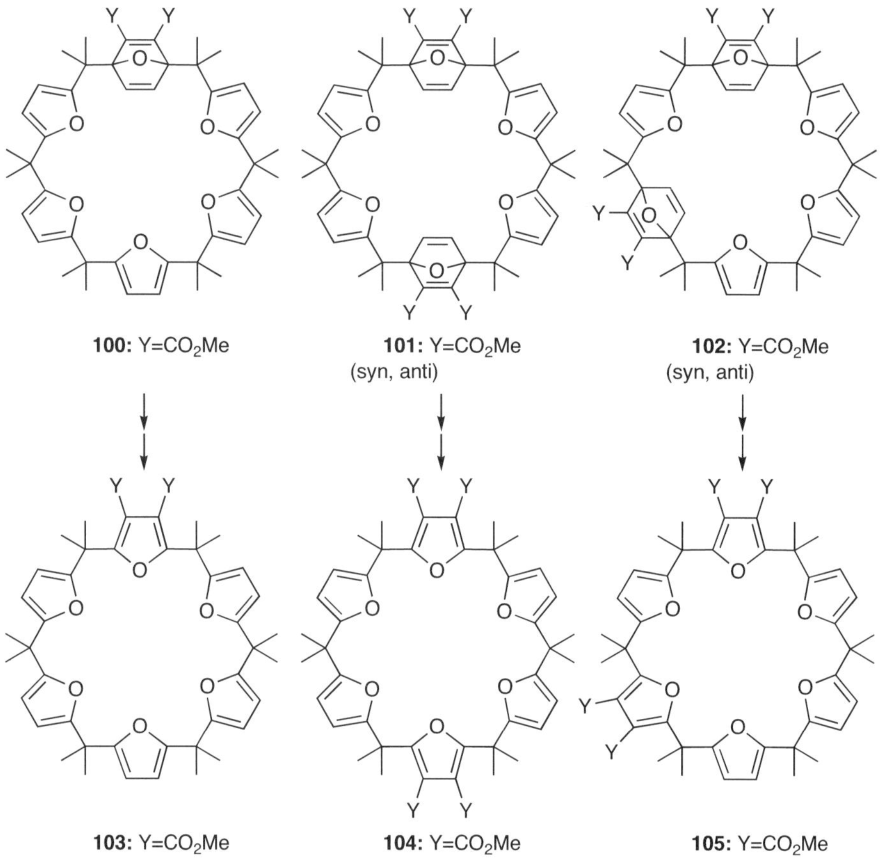

Scheme 14 Syntheses of substituted calix[6]furans via cycloadducts [53]

lution [3]. Heterocalix[4]arenes such as calix[4]furans and calix[4]pyrroles can also have these four typical conformations, and sometimes they show conformational preference different from that of calix[4]arenes.

It was reported that the parent calix[4]furan **4** crystallizes from ethanol in monoclinic and triclinic forms [22]. In the monoclinic form, **4** adopted

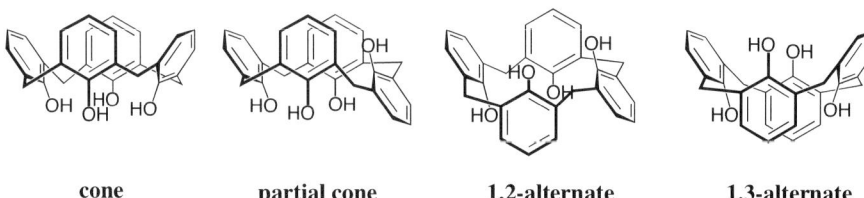

Fig. 1 Four typical conformations of calix[4]arene **106**

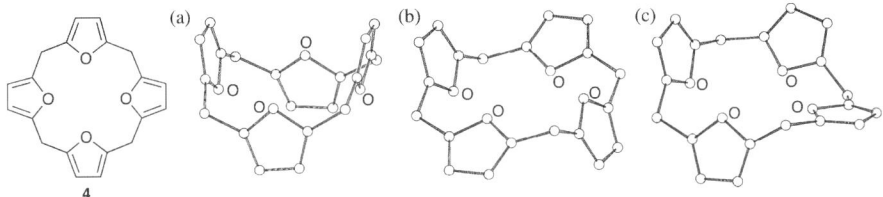

Fig. 2 Crystal structure of calix[4]furan **4** in monoclinic form (**a**), and two independent structures in triclinic form (**b** and **c**) [22]

a 1,3-alternate conformation with approximately D_{2d} symmetry (Fig. 2a). In another single crystal (triclinic), two conformations were observed (Fig. 2b and c), both of which differ significantly from that in the monoclinic form.

The DFT calculations (BLYP/6-31G**) on the four conformers of **4** indicated that the 1,3-alternate structure is the most stable, followed by the partial cone and the cone structures, while the 1,2-alternate structure is the least stable [54]. The calculated 1,3-alternate structure of **4** showed good agreement with the crystal structure observed in the monoclinic form. The calculations indicated that the 1,3-alternate conformation is also the most stable for the parent calix[4]pyrrole **34** and the parent calix[4]thiophene **84**. The preference of the 1,3-alternate conformation in these heterocalix[4]arenes was mainly explained in terms of the adjacent ring–ring electrostatic interaction. The calculated relative energies (BLYP/6-31G**) between *syn*- and *anti*-conformers of the substructures of heterocalix[4]arenes **4**, **34**, and **84** are shown in Fig. 3 (unit: kcal mol^{-1}). In all cases, the *anti* structure is more favorable than the *syn* structure. With the increase in the dipole moment of the heteroaromatic ring, from furan (0.91 D) to thiophene (0.98 D) and pyrrole (2.32 D), the energy difference becomes larger, and the cone conformation is more destabilized by the ring–ring electrostatic repulsion.

Calix[4]furan **1** with isopropylidene bridges also adopted the 1,3-alternate conformation, which is similar to that found in the monoclinic form of the parent calix[4]furan **4** [55]. The diameter of the central cavity of **1** is 2.02 Å, cf. 1.8 Å for 15-crown-5 and 2.8 Å for 18-crown-6 [56]. The crystal struc-

[kcal mol^{-1}]	*anti*	*syn*
X=O	0	0.5
X=S	0	1.3
X=NH	0	4.2

Fig. 3 Calculated relative energies (kcal mol^{-1}) between syn- and anti-conformations of substructures heterocalix[4]arenes (BLYP/6-31G**) [54]

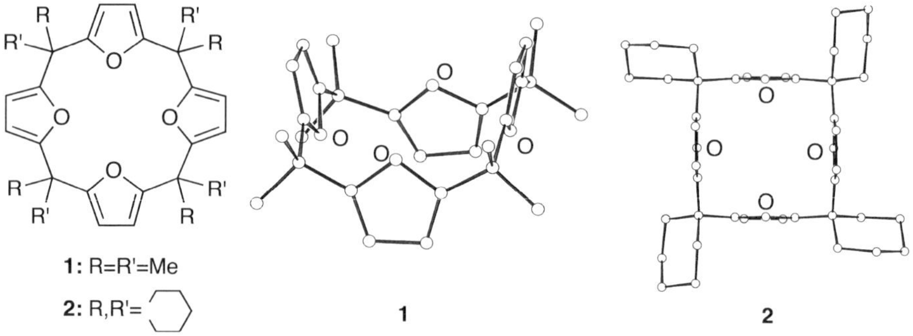

Fig. 4 Crystal structures of calix[4]furans **1** and **2** [55, 57]

ture of calix[4]furan **2** with cyclohexyl units showed a similar 1,3-alternate conformation [57]. In this structure, the furan rings of **2** are oriented in a nearly orthogonal up-down-up-down geometry. The angles between the mean plane of the macrocyclic ring and the furan-ring planes are 87–93°. These values are closer to orthogonal than those of calix[4]furans **4** (76–80° for monoclinic form) and **1** (77–80°). This tendency was attributed to the bulky cyclohexyl units interacting with the hydrogen atoms of the furan rings. The silicon-bridged calix[4]furan **8** also showed a 1,3-alternate structure in the crystal [27].

Among the larger family of calixfurans, only the crystallographic analysis of calix[6]furan **41** with isopropylidene bridges has been reported so far [40, 53]. The crystal structure of **41** has a crystallographic C_2 axis of symmetry passing through the two furan rings in the distal positions (Fig. 5). These two furan rings are oriented codirectionally and are approximately coplanar with the mean plane of the macrocycle. The other four furan rings adopt a near orthogonal up-down-up-down geometry. Such conformation of the macrocycle leaves almost no central void, despite the relatively large ring size.

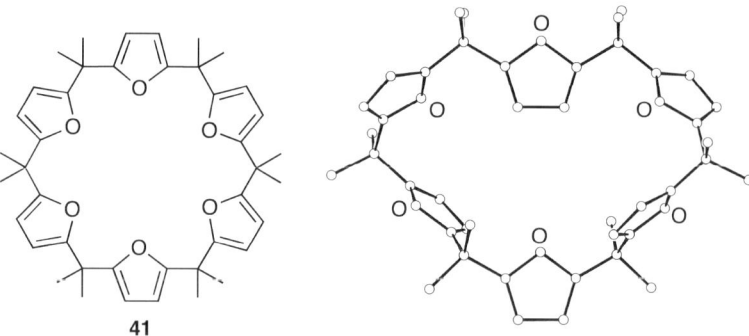

Fig. 5 Crystal structure of calix[6]furan **41** [40, 53]

4.2
Inclusion Properties of Calixfurans and Their Derivatives

Although the structures of calixfurans resemble those of crown ethers, their complexing abilities are generally poor due to the weak donor ability of the furan oxygen. Floriani et al. reported the synthesis of a unique silver complex of calix[4]furan **107** bearing ethyl groups on the bridges [58]. The reaction of **107** with silver triflate produced the dimeric complex, [**107**·AgSO$_3$CF$_3$]$_2$, the structure of which was characterized by X-ray analysis (Fig. 6). The equatorial plane around a pseudo-octahedral silver cation of the complex is determined by two oxygen atoms of the ligand **107** with a 1,3-alternate conformation and another two oxygen atoms of triflate anions. Interestingly, two axial coordination sites are filled by two C–H bonds, showing a (C–H)–M triangular arrangement, characteristic of a three center–two electron interaction. While there have been numerous examples of C–H agostic interactions, this complex presents an unprecedented type of side-on interaction of a C–H bond. It is probable that the poor binding ability of the calixfuran ligand plays a key role on the observation of such type of interaction.

In contrast with usual calixfurans with carbon bridges, the silicon-bridged calix[4]furan **8** was found to act as a good and specific chelator for Hg^{2+} ions, while its larger analogue, calix[6]furan **9**, did not show any significant metal binding ability [27]. A rationale for the metal affinity and specificity of **8** has not been given.

Compared with calixfurans, their saturated derivatives, i.e., calixtetrahydrofurans, such as **89** were found to show better binding abilities for metal cations [18, 20, 32]. Some calix[4]tetrahydrofuran derivatives have been employed as ionophores of lithium ion-selective membrane electrodes [34, 59].

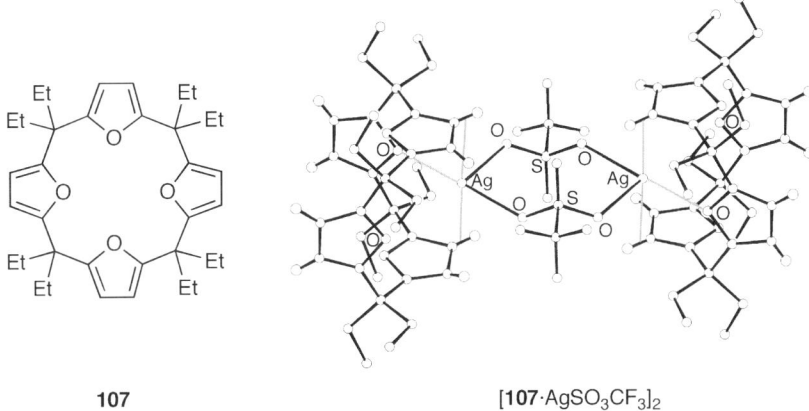

Fig. 6 Crystal structure of complex [**107**·AgSO$_3$CF$_3$]$_2$ [58]

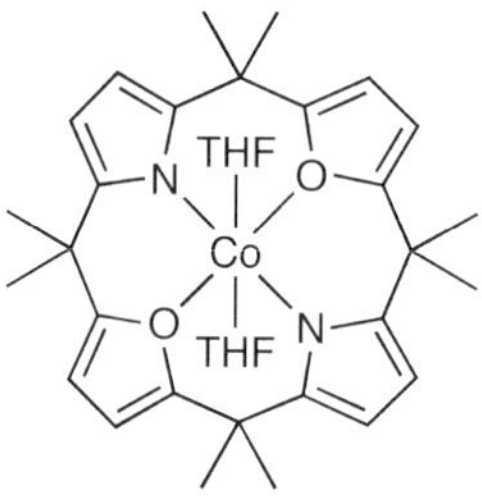

108

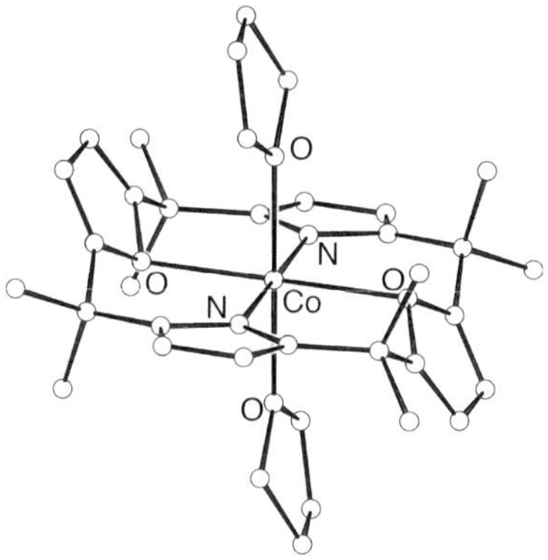

Fig. 7 Crystal structure of complex **108** [47]

Hybrid calixfuran systems containing other heterocycles have also been utilized as the ligands for metal cations. The dianionic ligand derived from calix[2]furan[2]pyrrole **76** forms the cobalt complex **108** in the reaction with $CoCl_2(THF)_{1.5}$, the structure of which was established by X-ray analysis (Fig. 7) [47]. The Co–O (furan) and Co–O (THF) bond lengths are 2.302(4) and 2.242(5) Å, respectively. There is a very weak Co–O interaction between the cobalt center and the furan oxygen, and the macrocyclic ligand set masks a dicoordinate cobalt in the cavity.

Some hybrid calixfurans containing pyrrole rings can serve as anion receptors. Kohnke et al. investigated the anion binding properties of hybrid calix[n]furan[m]pyrroles ($n + m = 6$) **80–82** as well as calix[6]pyrrole **79** [48]. As expected, the binding constants for halogen anions increased with the number of the pyrrole rings. The binding constants K_a (mol^{-1} dm^3) for Cl$^-$ (tetrabutylammonium salt) in wet CD_2Cl_2 are $1.2 \times 10^4 \pm 10^3$ for **79**, $5.5 \times 10^3 \pm 600$ for **80**, and 65 ± 8 for **81** while no detectable binding was observed for **82**. Calix[1]furan[5]pyrrole **80** and calix[2]furan[4]pyrrole **81** showed complexing ability for chloride and fluoride anions, respectively, whereas calix[3]furan[3]pyrrole **82** did not undergo a detectable complexation with halogen anions. The structure of the complex [**80**·Cl$^-$][18-crown-6·K$^+$] is shown in Fig. 8. The binding constant of the hybrid cyclic hexamer **81**, containing four pyrrole units and two furan units, with fluoride is larger than that of calix[4]pyrrole **77**, indicating the favorable effect of a larger cavity size when the number of pyrrole units is kept constant. It is also notable that **81** exhibits considerably greater selectivity between fluoride and chloride than **77**.

The strapped calix[2]furan[4]pyrrole **63** with a cryptand-like structure showed moderate binding ability for fluoride. It was suggested that **63** can possibly bind CsF as an ion pair in the cavity [44].

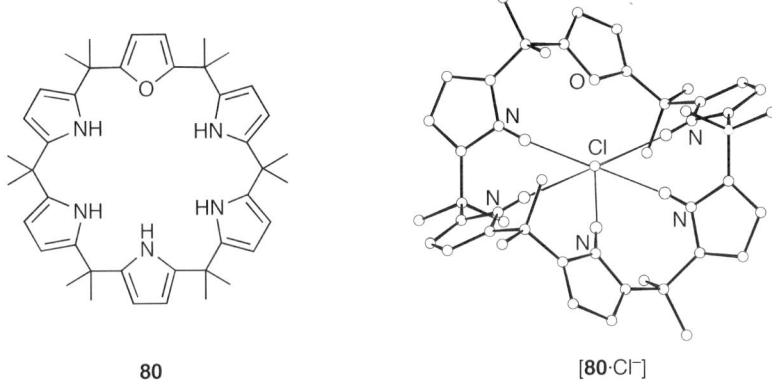

Fig. 8 Crystal structure of complex [80·Cl⁻] [48]

5
Concluding Remarks

Calixfurans appear to be a tactful supporting actor in the chemistry of calixarenes. Although their intrinsic binding abilities are rather modest, the weak coordination by the furan units of calixfurans or hybrid systems plays a crucial role in some cases. More importantly, calixfurans can transform themselves into a wide variety of macrocycles including those otherwise difficult to access. Further development of the synthetic strategy of calixfurans as well as the novel methods for their transformation to other functional molecules is expected. Since conformational behavior of calixfurans has not been sufficiently clarified yet, the more sophisticated strategy for regulation of their conformational dynamics should be further explored for the ready construction of the desired molecular framework.

References

1. Gutsche CD (1989) In: Stoddart JF (ed) Calixarenes. Royal Society of Chemistry, Cambridge
2. Vicens J, Bömer V (eds) (1991) Calixarenes: a versatile class of macrocyclic compounds. Kluwer, Dordrecht
3. Gutsche CD (1998) In: Stoddart JF (ed) Calixarenes revisited. Royal Society of Chemistry, Cambridge
4. Pochini A, Arduini A (2000) In: Mandolini L (ed) Calixarenes in action. Imperial College Press, London
5. Asfari Z, Bömer V, Harrowfield J, Vicens J, Saadioui M (eds) (2001) Calixarenes 2001. Kluwer, Dordrecht
6. Newkome GR, Sauer JD, Roper JM, Hager DC (1977) Chem Rev 77:513
7. Ibach S, Prautzsch V, Vögtle F, Chartroux C, Gloe K (1999) Acc Chem Res 32:729

8. Sliwa W (2004) Chem Heterocycl Compd 40:683
9. Kumar S, Paul D, Singh H (2006) Adv Heterocycl Chem 89:65
10. Gale PA, Sessler JL, Kral V (1998) Chem Commun 1
11. Sessler JL, Gale PA (2000) In: Kadish KM, Smith KM, Guilard R (eds) Porphyrin handbook. Academic Press, New York
12. Gale PA, Anzenbacher P Jr, Sessler JL (2001) Coord Chem Rev 222:57
13. Meyers AI (1974) Heterocycles in organic synthesis. Wiley-Interscience, New York
14. Butin AV, Stroganova TA, Kulnevich VG (1999) Chem Heterocycl Chem 35:757
15. Hale WJ, McNally WD, Pater CJ (1906) Am Chem J 35:72
16. Ackman RG, Brown WH, Wright GF (1955) J Org Chem 20:1147
17. Beals RE, Brown WH (1956) J Org Chem 21:447
18. Chastrette M, Chastrette F (1973) J Chem Soc Chem Commun 534
19. Healy MdS, Rest AJ (1981) J Chem Soc Chem Commun 149
20. Healy MdS, Rest AJ (1985) J Chem Soc Perkin Trans 1 973
21. Pajewski R, Ostaszewski R, Jurczak J (2000) Org Prep Proced Int 32:394
22. Haas W, Knipp B, Sicken M, Lex J, Vogel E (1988) Angew Chem Int Ed 27:409
23. Vogel E, Roehrig P, Sicken M, Knipp B, Herrmann A, Pohl M, Schmickler H, Lex J (1989) Angew Chem Int Ed 28:1651
24. Musau RM, Whiting A (1994) J Chem Soc, Perkin Trans 1 2881
25. Vogel E, Doerr J, Herrmann A, Lex J, Schmickler H, Walgenbach P, Gisselbrecht JP, Gross M (1993) Angew Chem Int Ed 32:1597
26. König B, Rödel M, Bubenitschek P, Jones PG (1995) Angew Chem Int Ed 34:661
27. König B, Rödel M, Bubenitschek P, Jones PG, Thondorf I (1995) J Org Chem 60:7406
28. Brown WH, Hutchinson BJ, MacKinnon MH (1971) Can J Chem 49:4017
29. Pajewski R, Ostaszewski R, Ziach K, Kulesza A, Jurczak J (2004) Synthesis 865
30. Brown WH, French WN (1958) Can J Chem 36:537
31. Brown WH, Hutchinson BJ (1978) Can J Chem 56:617
32. Kobuke Y, Hanji K, Horiguchi K, Asada M, Nakayama Y, Furukawa J (1976) J Am Chem Soc 98:7414
33. Hoegberg AGS, Weber M (1983) Acta Chem Scand B37:55
34. Kang YR, Lee KM, Nam H, Cha GS, Jung SO, Kim JS (1997) Analyst 122:1445
35. Brown WH, French WN (1958) Can J Chem 36:371
36. Song M-Y, Na H-K, Kim E-Y, Lee S-J, Kim KI, Baek E-M, Kim H-S, An DK, Lee C-H (2004) Tetrahedron Lett 45:299
37. Jang Y-S, Kim H-J, Lee P-H, Lee C-H (2000) Tetrahedron Lett 41:2919
38. Nagarajan A, Ka JW, Lee CH (2001) Tetrahedron 57:7323
39. Taniguchi S, Hasegawa H, Yanagiya S, Tabeta Y, Nakano Y, Takahashi M (2001) Tetrahedron 57:2103
40. Kohnke FH, La Torre GL, Parisi MF, Menzer S, Williams DJ (1996) Tetrahedron Lett 37:4593
41. Beneteau V, Meth-Cohn O, Rees CW (2001) J Chem Soc Perkin Trans 1 3297
42. Cafeo G, Kohnke FH, La Torre GL, White AJP, Williams DJ (2000) Angew Chem Int Ed 39:1496
43. Arumugam N, Jang Y-S, Lee C-H (2000) Org Lett 2:3115
44. Cafeo G, Kaledkowski A, Kohnke FH, Messina A (2006) Supramol Chem 18:273
45. Williams PD, LeGoff E (1981) J Org Chem 46:4143
46. Cafeo G, Kohnke FH, Parisi MF, Nascone RP, La Torre GL, Williams DJ (2002) Org Lett 4:2695
47. Crescenzi R, Solari E, Floriani C, Chiesi-Villa A, Rizzoli C (1996) Inorg Chem 35:2413

48. Cafeo G, Kohnke FH, La Torre GL, Parisi MF, Nascone RP, White AJP, Williams DJ (2002) Chem Eur J 8:3148
49. Cafeo G, Garozzo D, Kohnke FH, Pappalardo S, Parisi MF, Nascone RP, Williams DJ (2004) Tetrahedron 60:1895
50. Guillard J, Meth-Cohn O, Rees CW, White AJP, Williams DJ (2002) Chem Commun 232
51. Vogel E, Haas W, Knipp B, Lex J, Schmickler H (1988) Angew Chem Int Ed 27:406
52. Hart H, Takehira Y (1982) J Org Chem 47:4370
53. Cafeo G, Giannetto M, Kohnke FH, La Torre GL, Parisi MF, Menzer S, White AJP, Williams DJ (1999) Chem Eur J 5:356
54. Wang D-F, Wu Y-D (2004) J Theor Comput Chem 3:51
55. Hazell A (1989) Acta Crystallogr, Sect C: Cryst Struct Commun C45:137
56. Lamb JD, Izatt RM, Christensen JJ, Eatough DJ (1979) In: Melson GA (ed) Coordination chemistry of macrocyclic compounds. Plenum, New York, London
57. Pajewski R, Pecak A, Ostaszewski R, Jurczak J (1999) Acta Crystallogr, Sect C: Cryst Struct Commun C55:1862
58. Kretz CM, Gallo E, Solari E, Floriani C, Chiesi-Villa A, Rizzoli C (1994) J Am Chem Soc 116:10775
59. Kim JS, Jung SO, Lee SS, Kim SJ (1993) Bull Korean Chem Soc 14:123

Supramolecules Based on Porphyrins

Hiroko Yamada[1,2] (✉) · Tetsuo Okujima[1] · Noboru Ono[1]

[1]Department of Chemistry and Biology, Graduate School of Science and Engineering, Ehime University, 2-5 Bunkyo-cho, 790-8577 Matsuyama, Japan
yamada@chem.sci.ehime-u.ac.jp

[2]PRESTO, JST, 332-0012 Kawaguchi, Japan

1	Introduction	124
1.1	General Aspects of Supramolecules of Porphyrins and Phthalocyanines	124
1.2	Metal–Ligand Coordinative Interactions	126
1.3	Hydrogen-Bonding Interactions	127
2	Optical Applications	131
2.1	Two-Photon Absorption (TPA)	131
2.2	Porphyrin Monomers and Covalently Linked Oligomers for TPA	131
2.3	Supramolecules of Porphyrin Oligomers for TPA	133
3	Artificial Photosynthesis	137
3.1	Dye-Sensitized Solar Cells and Thin-Film Organic Photovoltaic Cells	137
3.2	Donor–Acceptor Systems by π–π Interactions	139
3.3	Photoconversion Systems by Coordination, Hydrogen Bonding and Electrostatic Interactions	143
4	Electronic Applications	148
4.1	General Aspects of Organic Field-Effect Transistors (OFETs)	148
4.2	OFETs by Solution-Processed Tetrabenzoporphyrins (TBPs)	151
5	Concluding Remarks	155
	References	155

Abstract As porphyrins and phthalocyanines possess unique electronic, magnetic and optical properties, supramolecular assembly based on them is subject to intense research targets. Herein, the reviewers focus on the supramolecular architectures of porphyrins, which enable their use as electronic and optical functional materials such as third-order optical susceptibilities, photoenergy conversion systems, and organic field-effect transistors.

Keywords Optical and electronic devices · π–π stacking interaction · Porphyrin · van der Waals interaction

Abbreviations
A   Acceptor
AFM   Atomic force microscopy
a-Si   Amorphous silicon

BCOD   Bicyclo[2.2.2]octadiene
BCP   2,9-dimethyl-4,7-diphenyl-1,10-phenanthroline
CP   BCOD-fused porphyrin
CT   Charge transfer
D   Donor
DNA   Deoxyribonucleic acid
DSSC   Dye-sensitized solar cell
EDOT   3,4-ethylenedioxythiophene
Fc   Ferrocene
$F_{16}Pc$ Fluorinated phthalocyanine
IPCE   Incident photon to photocurrent efficiency
ITO   Indium tin oxide
LB   Langmuir–Blodgett
NLO   Nonlinear optics
OEP   2,3,7,8,12,13,17,18-octaethylporphyrin
OFET   Organic field-effect transistor
OLED   Organic light emitting diode
Pc   Phthalocyanine
PCBM   [6,6]-phenyl-C_{61}-butyric acid methyl ester
PCBNB   [6,6]-phenyl-C_{61}-butyric acid *n*-butyl ester
PEDOT   Poly(3,4-ehylenedioxythiophene)
PSS   Poly(styrene sulfonate)
Py   Pyridyl group
PyP   *meso*-pyridyl porphyrin
SAM   Self-assembled monolayer
SWNT   Single-wall carbon nanotube
TBP   Tetrabenzoporphyrin
TPA   Two-photon absorption
TPFP   Tetra(pentafluorophenyl)porphyrin
TPP   Tetraphenylporphyrin
V_{th} Threshold voltage
XRD   X-ray diffraction

1
Introduction

1.1
General Aspects of Supramolecules of Porphyrins and Phthalocyanines

Control of molecular self-assembly to generate supramolecular architectures that are organized in well-defined geometries is important in various fields of science and technology [1, 2]. As porphyrins and phthalocyanines (Pcs) serve as components of molecular materials that possess unique electronic, magnetic and optical properties [3, 4], supramolecular assembly based on them is subject to intense research targets. In this section the basic elements of supramolecules based on porphyrins are described and the creation of optical and electronic functional properties will be discussed in the following sections.

Porphyrins and Pcs tend to align into one-dimensional aggregates to create supramolecular architectures such as nanowires, discotic liquid crystals, helical ribbon structures, etc. The major driving forces are considered to be $\pi-\pi$ stacking and/or van der Waals interaction [3]. Our daily life benefits from such supramolecular architectures of porphyrins and Pcs. Pcs as aggregates in the solid state or thin films absorb light in near red regions (600–850 nm) and are practically used as photoconductors. In particular, PcTiO is useful as a charge generation material for laser printers. The photocarrier efficiency depends not only on the central metal atoms but also on the crystal structure [5].

In nature, chlorophylls self-aggregate to form the main light-harvesting antennae of photosynthetic green bacteria [6]. Tetrapyrrole dyes are placed in a precise and specific arrangement as found in photosynthesis, where the proteins serve as templates. For example, circularly arranged chromophoric assemblies are found in the light-harvesting complex employed for photosynthesis by green plants and purple bacteria [7]. Since its discovery, many efforts have been directed towards the synthesis of cyclic porphyrin arrays to study excitation energy transfer along the cyclic arrays [8, 9]. In order to construct such cyclic porphyrin arrays, a supramolecular approach is attractive. For example, Zn chlorophyll with a *meso*-pyridyl group (Py) **1** is self assembled to form tetramers, where ultrafast energy transfer within cyclic tetramers is observed (Fig. 1) [10].

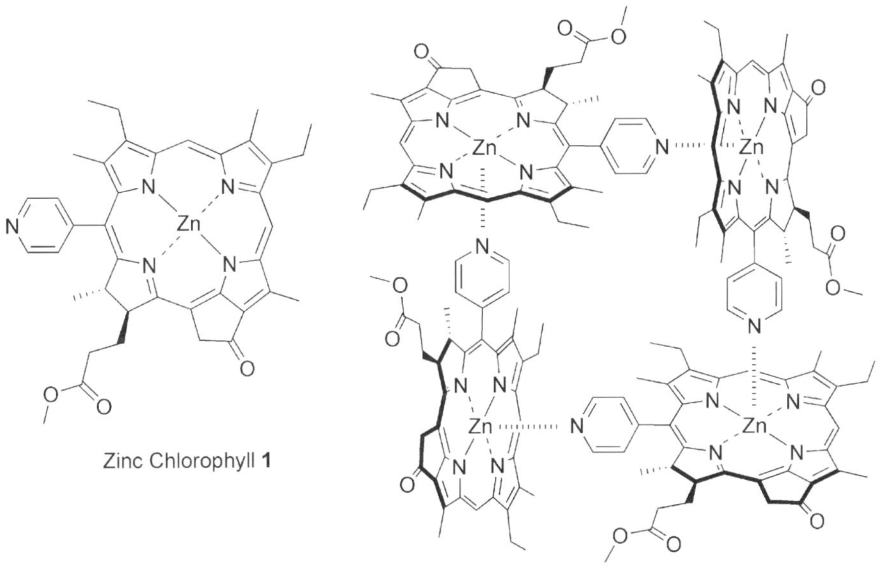

Fig. 1 Structure of Zn chlorophyll **1** and its tetramer [10]

1.2
Metal–Ligand Coordinative Interactions

Porphyrin assemblies induced by metal–ligand coordinative interactions have been the subject of numerous studies in suprachemistry. In order to organize porphyrin chromophores into well-defined supramolecular architectures, various classes of peripheral substituents are introduced as coordination sites for metalloporphyrins. Pyridyl, imidazolyl, aminopyrimidyl, amino, and phosphoryl groups have been used as ligands to construct cofacial, linear, branched, cyclic, dendric and polymeric metalloporphyrins [11–15]. Among them, *meso*-pyridyl porphyrins (PyPs) have been most widely used to construct geometrically well-defined molecular assemblies by coordination of the pyridyl groups. The peripheral N atom of PyPs can be in either the 4′- or 3′-position. In general, the exocyclic coordination bonds are established in the plane of the porphyrins with 4′-PyPs, and with 3′-PyPs, the coordination bonds are directed out of the plane of the porphyrin as shown in Fig. 2 [16]. Pyridine–zinc ligation provides a useful tool for template-directed synthesis of cyclic porphyrin oligomers. Carefully designed templates containing pyridines with a proper symmetry afford porphyrin oligomers in good yields. Two recent examples of template-directed synthesis of porphyrin nanowires are presented in Figs. 3 and 4 [17, 18].

Self organization induced by imidazolyl groups affords slipped cofacial dimers. Kobuke et al. have used this strategy to construct assemblies of slipped cofacial porphyrins **6** showing strong electronic coupling (Fig. 5) [19, 20]. Kameyama et al. have extended this strategy to metallophthalocyanines to obtain highly fluorescent self-coordinated phthalocyanine dimers **7** [21].

The Pd-catalyzed amination reaction of *meso*-hexynyl Zn(II) porphyrin with 4-amino-3-iodopyridine provides *meso*-(5-azaindolyl)-substituted Zn(II) porphyrin **8**, which assembles to form a slipped cofacial dimer by the coordination of the pyridine moiety to the Zn(II) center (Fig. 6) [22]. The condensation reaction of *meso*-amino Zn(II) porphyrin with cinchomeronic anhydride affords *meso*-cinchomeronimide-substituted Zn(II) porphyrin **9**, which forms a cyclic trimer and its *meso–meso* linked Zn(II) diporphyrins assemble to form a discrete cyclic trimer, tetramer, and pentamer (Fig. 7) [23].

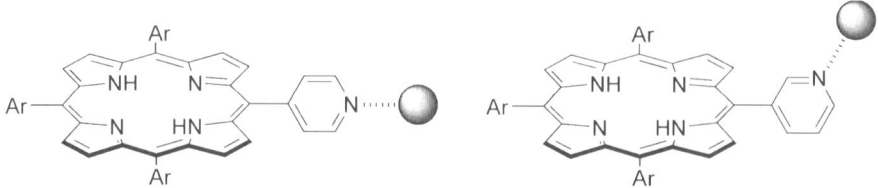

Fig. 2 Coordination interaction of *meso*-Py and metal

Fig. 3 Template-directed synthesis of porphyrin nanoring **3** [17]

Another approach to such a slipped dimer is self organization of *meso*-phosphorylporphyrins **10** through P-oxo-Zinc coordination (Fig. 8) [24].

1.3
Hydrogen-Bonding Interactions

Hydrogen bonding is also important to construct supramolecular architectures of porphyrins, which is well summarized in the review by Satake and Kobuke [14]. For example, self-assembly of a porphyrin–fullerene dyad through Watson–Click hydrogen bonding offers a good model of photoinduced electron transfer in supramolecular assembly (Fig. 9) [25]. Shirakawa

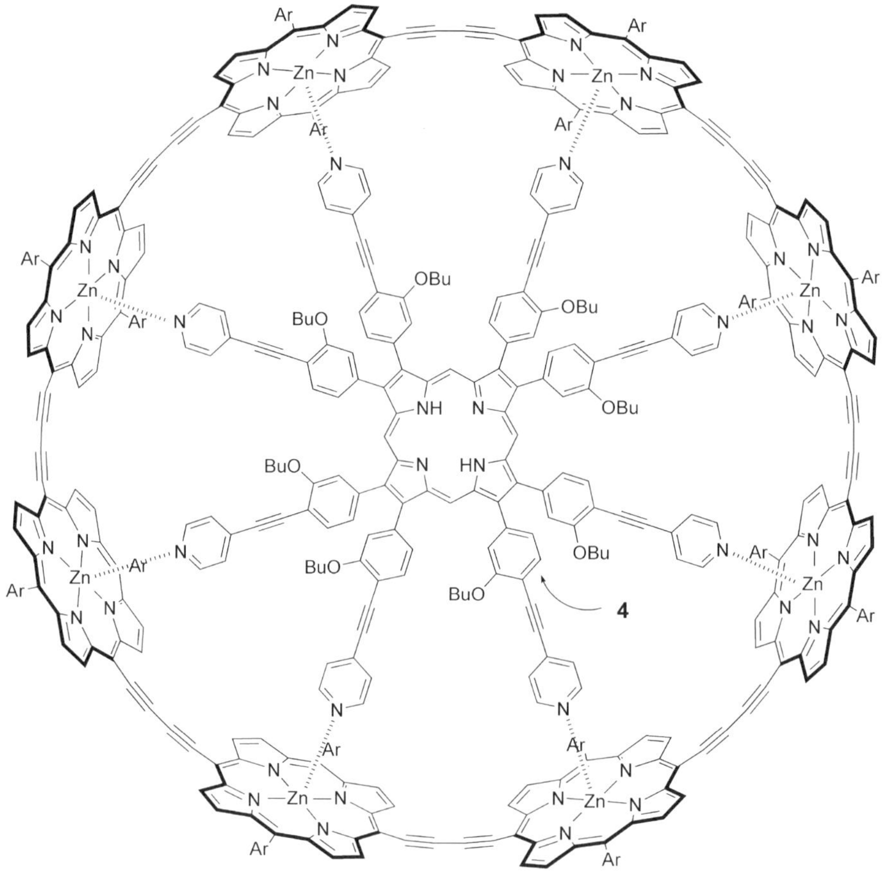

Fig. 4 Template-directed synthesis of porphyrin nanoring 5 [18]

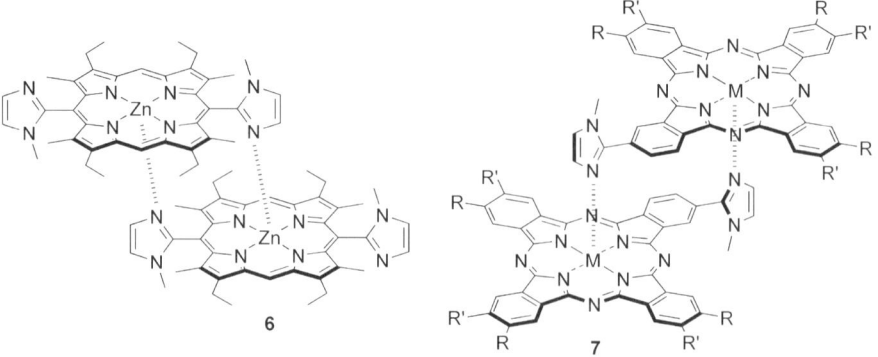

Fig. 5 Self-coordinated dimers of porphyrin 6 and phthalocyanine 7 [19, 21]

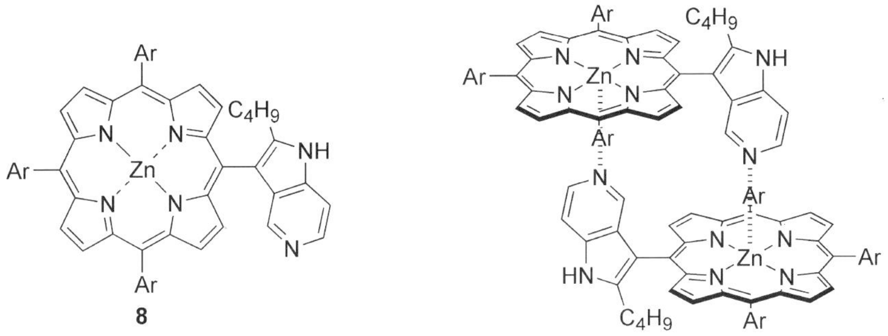

Fig. 6 Dimeric assembly of porphyrin **8** [22]

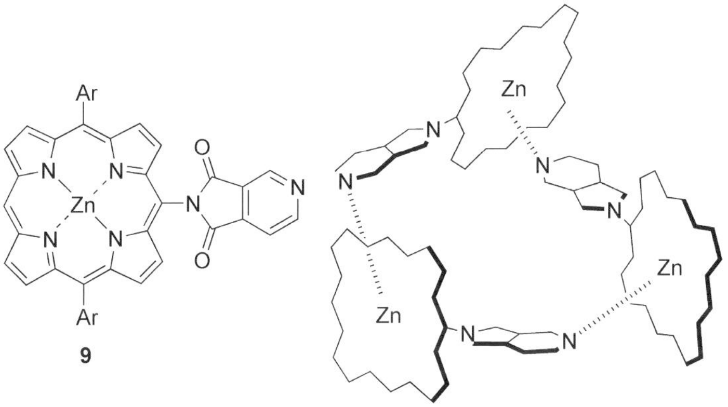

Fig. 7 Cyclic trimer of porphyrin **9** [23]

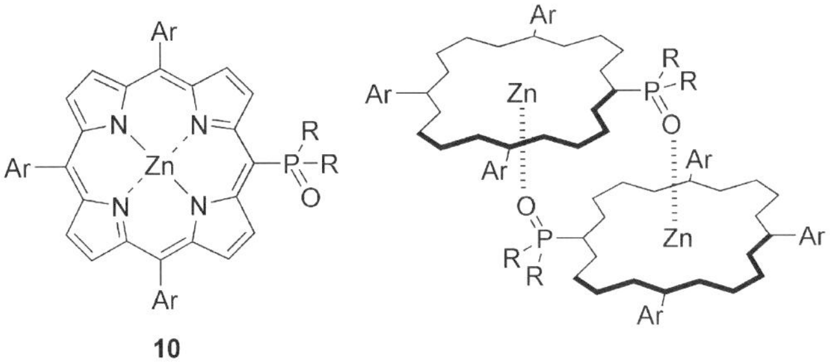

Fig. 8 P–O–Zn coordination of porphyrin **10** [24]

et al. reported on the interesting hydrogen bond-assisted control of the *J*- vs. *H*-aggregation mode of porphyrin-stacks in an organogel system. Porphyrins with amide groups as peripheral hydrogen bonding sites act as a gelator

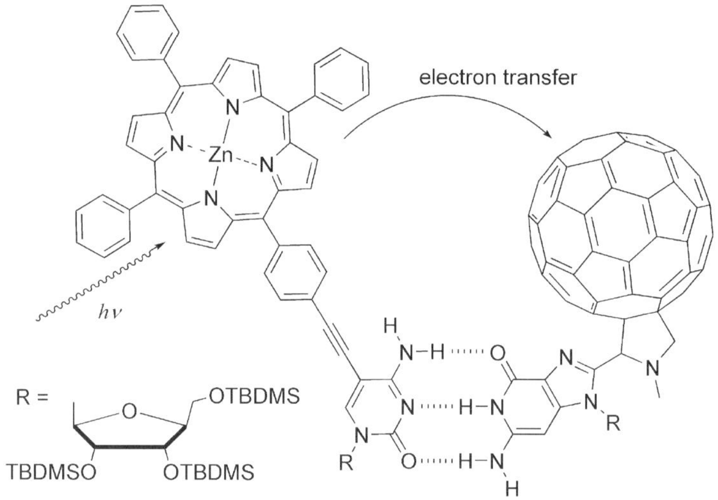

Fig. 9 Self-assembly of porphyrin–fullerene dyad [25]

forming a one-dimensional aggregate; a J-aggregate by aggregation of porphyrin **11** or an H-aggregate by porphyrin **12**, depending on the positions of the amide groups [26]. In the presence of C_{60}, porphyrin **11b** assembled to form a one-dimensional $(C_{60})_n(\mathbf{11b})_{2n}$ aggregate as illustrated in Fig. 10 [27, 28].

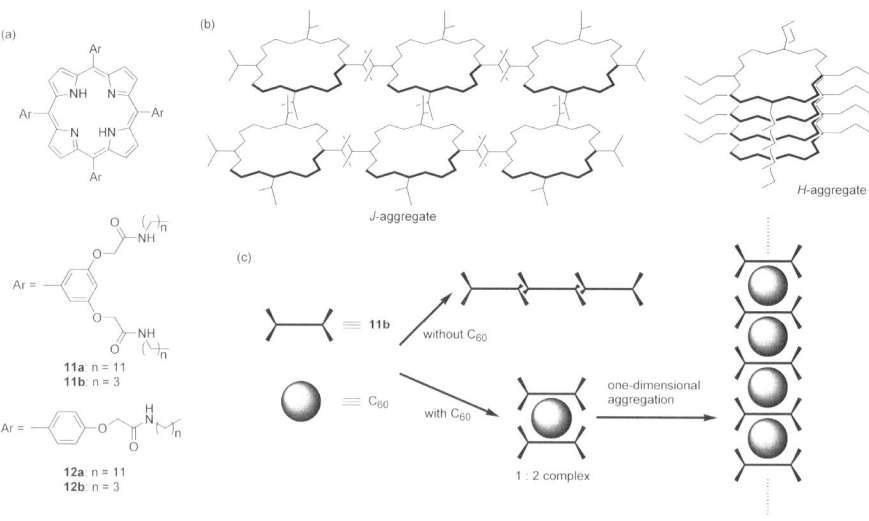

Fig. 10 a Structure of amino-appended porphyrins **11** and **12**; **b** J- and H-aggregate of porphyrins **11** and **12**, respectively, and **c** one-dimensional aggregate of C_{60} and **11b** [26, 27]

2
Optical Applications

2.1
Two-Photon Absorption (TPA)

In this section, porphyrin supramolecules indicating strong third-order optical susceptibility are discussed. Materials with large third-order optical susceptibility have numerous applications in nonlinear optics, such as ultrafast optical switching and modulations. Especially two-photon absorption (TPA) has been focused on due to the large number of potential applications, such as photodynamic therapy, optical power limiting, three-dimensional microfabrication, two-photon-excited fluorescence spectroscopy, and so on. TPA is a phenomenon where excitation occurs by the simultaneous absorption of two photons at wavelength 2λ using strong laser pulses, instead of a single-photon excitation at a wavelength of λ. TPA efficiency is quantified by the two-photon absorption cross-section $\sigma^{(2)}$ in GM, which corresponds to 10^{-50} cm^4 sec^{-1} molecule^{-1} photon^{-1}. Strong third-order nonlinearity is exhibited by materials with highly π-conjugated systems. Hence, porphyrins and Pcs are good candidates for nonlinear optical materials. As Pcs have remarkable chemical and thermal stability, they are ideal materials for nonlinear optics (NLO). O'Flaherty et al. have provided a comprehensive review on Pcs for optical limiting and NLO [29]. Covalently linked or self-assembled porphyrin array systems with enhanced TPA have also been reported [30]. Here porphyrins with large third-order optical nonlinearity are focused upon.

2.2
Porphyrin Monomers and Covalently Linked Oligomers for TPA

Before proceeding to a discussion of supramolecules of porphyrins we will provide a brief description of porphyrin monomers and connected dimers as materials for TPA. Several π-conjugation expanded acetylene- or butadiyne-linked porphyrin dimers have been prepared and their TPA properties have been investigated (Fig. 11) [31–33]. Porphyrin **13** shows 400-fold larger intrinsic (femtosecond) TPA cross sections (6.0×10^3 GM), compared to the parent monomer. Ahn et al. controlled the dihedral angles of directly linked porphyrin dimers and arrays **14–18** to explore the relationship between the π-conjugation effect of the adjacent porphyrin planes and the TPA values (Fig. 12) [34]. The enhancement in electronic interactions in various porphyrin arrays upon reduction of the dihedral angle is strongly correlated with the large TPA cross section. The same group have reported triply linked porphyrin dimer, trimer, and tetramer **19** ($n = 2$–4) which exhibits larger TPA cross sections (1.2×10^4–9.4×10^4 GM) compared to the porphyrin monomers owing to much larger TPA cross section values induced by nearly complete π-electron delocalization without interruption throughout the whole array skeleton (Fig. 13) [34, 35]. Triply linked dibenzoporphyrin dimer **20** also shows a large TPA cross section (1.5×10^4 GM) because of its expanded π-conjugation by peripheral benzene rings [36].

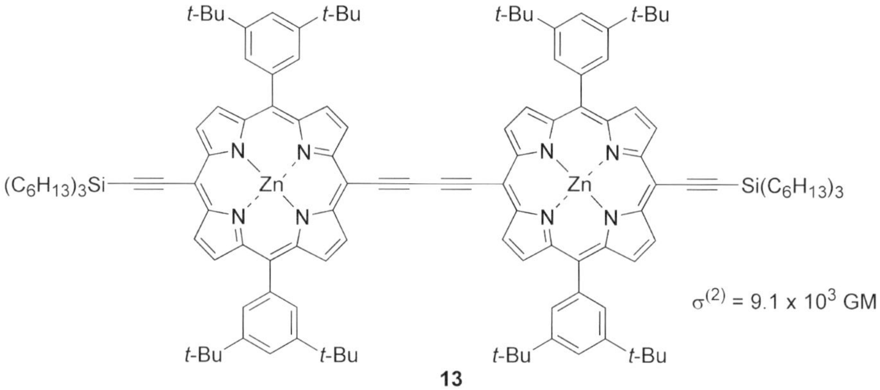

Fig. 11 Molecular structure of butadiyne-linked porphyrin dimer **13** [31–33]

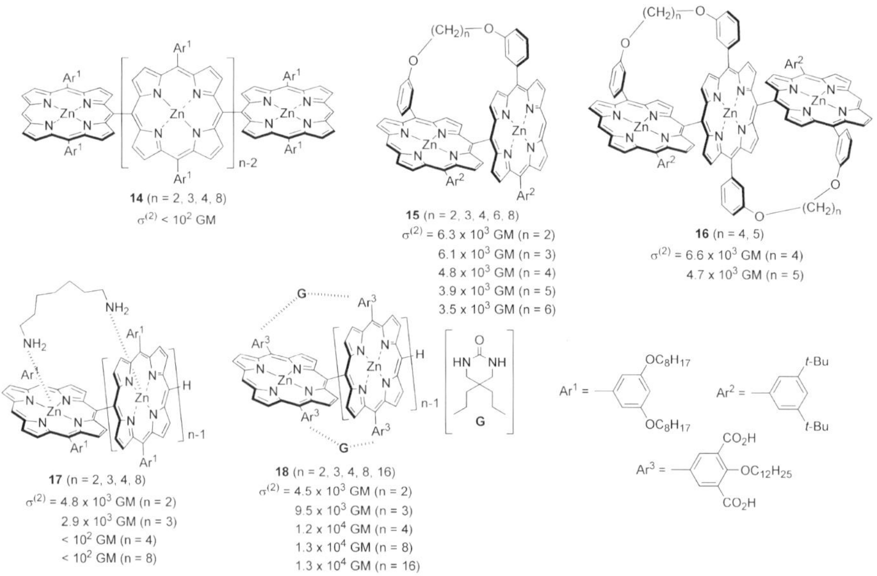

Fig. 12 Molecular structure of dihedral angle controlled, directly linked porphyrin dimers and arrays **14–18** [34]

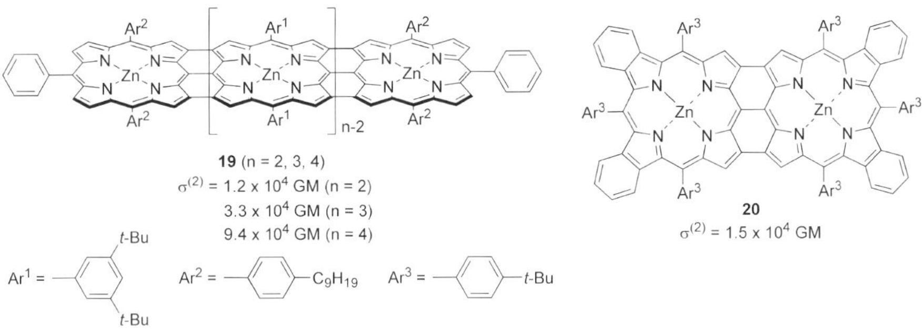

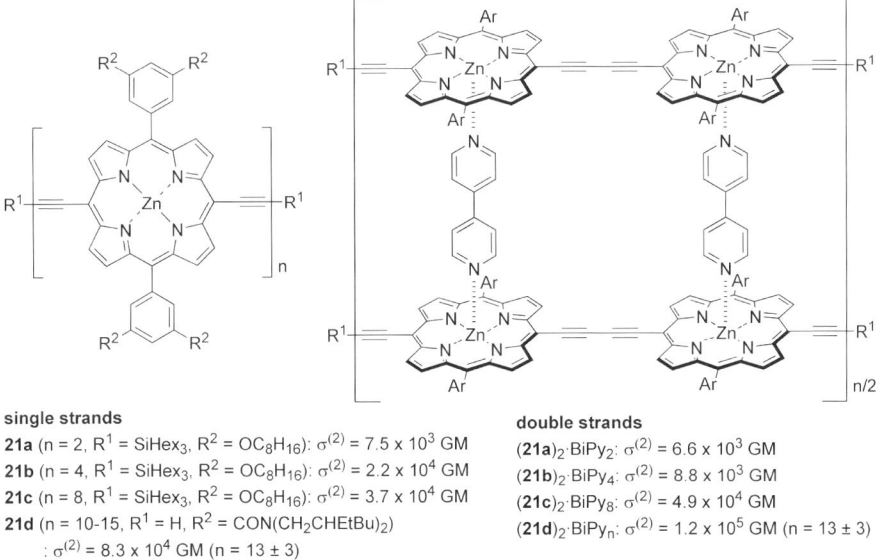

Fig. 13 Molecular structure of triply linked porphyrin dimer, trimer, tetramer **19**, and benzoporphyrin dimer **20** [35, 36]

2.3 Supramolecules of Porphyrin Oligomers for TPA

Linearly expanded porphyrin oligomers and polymers with enhanced third-order susceptibilities have also been reported. In solution the conjugation lengths of single-strand chains can be strongly limited by rotation and disorder. The self-assembly of double-strand ladder complexes of porphyrin butadiyne oligomers **21** achieves amplification in the nonlinearity

single strands

21a (n = 2, R^1 = $SiHex_3$, R^2 = OC_8H_{16}): $\sigma^{(2)} = 7.5 \times 10^3$ GM
21b (n = 4, R^1 = $SiHex_3$, R^2 = OC_8H_{16}): $\sigma^{(2)} = 2.2 \times 10^4$ GM
21c (n = 8, R^1 = $SiHex_3$, R^2 = OC_8H_{16}): $\sigma^{(2)} = 3.7 \times 10^4$ GM
21d (n = 10-15, R^1 = H, R^2 = $CON(CH_2CHEtBu)_2$) : $\sigma^{(2)} = 8.3 \times 10^4$ GM (n = 13 ± 3)

double strands

$(\mathbf{21a})_2 \cdot BiPy_2$: $\sigma^{(2)} = 6.6 \times 10^3$ GM
$(\mathbf{21b})_2 \cdot BiPy_4$: $\sigma^{(2)} = 8.8 \times 10^3$ GM
$(\mathbf{21c})_2 \cdot BiPy_8$: $\sigma^{(2)} = 4.9 \times 10^4$ GM
$(\mathbf{21d})_2 \cdot BiPy_n$: $\sigma^{(2)} = 1.2 \times 10^5$ GM (n = 13 ± 3)

Fig. 14 Molecular structure of single- and double-strand butadiyne-linked porphyrins **21** [37, 38]

of these polymers by holding the π-systems in the planar conformation (Fig. 14) [37, 38]. For example, the TPA cross section of butadiyne-linked porphyrin polymer **21d** achieves amplification by self-assembly of a double-strand ladder [(**21d**)$_2$·BiPy$_n$, 1.15×10^5 GM] from the single strand (**21d**, 8.3×10^4 GM) [38]. The TPA maximum peaks have red-shifted about 340 nm upon transition from a single- to double-strand structure for butadiyne-linked oligomers ($n = 4, 8$, and 13).

To enhance the TPA cross section, molecular designs involving donor/acceptor (D/A) sets intervened with a π-conjugation system in a symmetrical (D–π–D or A–π–A) or asymmetrical arrangement (D–π–A) have been proposed (Fig. 15). Ogawa et al. have investigated the A–π–A system using self-assembled porphyrin tetramer **22** of butadiyne-linked zinc-free base porphyrin dimers, with a TPA cross section of 7.6×10^3 GM by excitation with a 120 fs pulse and 2.1×10^5 GM with a 5 ns pulse. The TPA cross section of self-assembled porphyrin tetramer **23** of *meso–meso* linked porphyrin dimers (3.7×10^2 GM with the 120 fs pulse) is 20-times smaller than that of porphyrin tetramer **22**. It has been revealed that for a larger TPA cross section, a coplanar orientation like the butadiyne-linked dimer is more profitable than the orthogonal conformation of a *meso–meso* linked porphyrin dimer [39, 40]. The self-assembled polymer **24** of a butadiene-linked zinc porphyrin dimer shows 4.4×10^5 GM

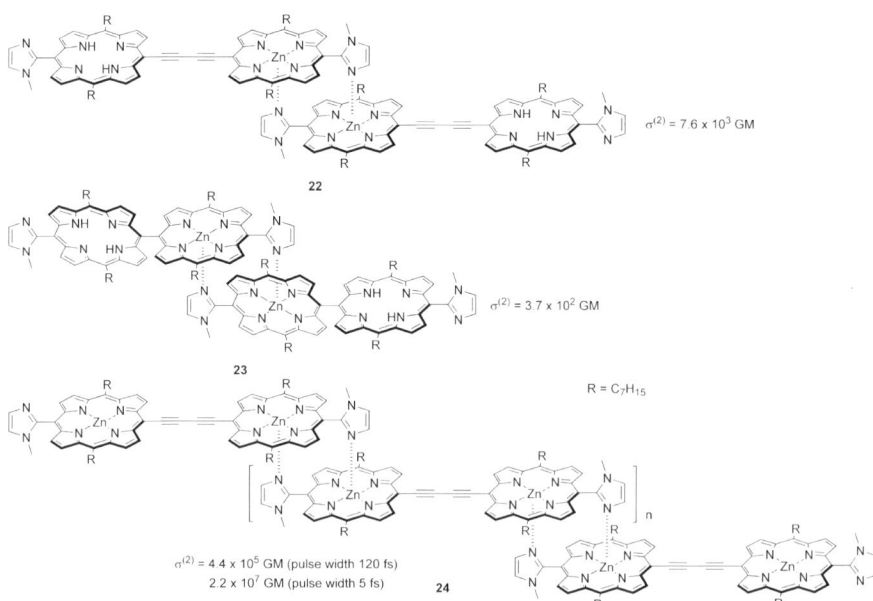

Fig. 15 Self-assembled porphyrin arrays **22–24** by complementary coordination of imidazole moiety to zinc [39, 40]

with a 120 fs pulse and 2.2×10^7 GM with a 5 ns pulse (Fig. 15) [40]. When ferrocene (D) or acceptors (A) or both (D/A) are placed at the end of the tetramers (**25**, **26**, and **27** for D–π–D, A–π–A, and D–π–A, respectively), the order of TPA cross sections is D–π–A (2.0×10^5 GM) > A–π–A (1.7×10^5 GM) > D–π–D (1.5×10^5 GM) (Fig. 16) [41]. The results suggest that the asymmetric D–π–A structure is advantageous for enhancing TPA in this series of molecules. Recently, self-assembled water-soluble porphyrin tetramers **28** and **29** have been reported to show 7.9×10^3 and 3.3×10^4 GM, respectively, with a 5 ns pulse at 870 nm in aqueous solution (Fig. 17) [42, 43]. These porphyrins might be potential candidates for TPA-

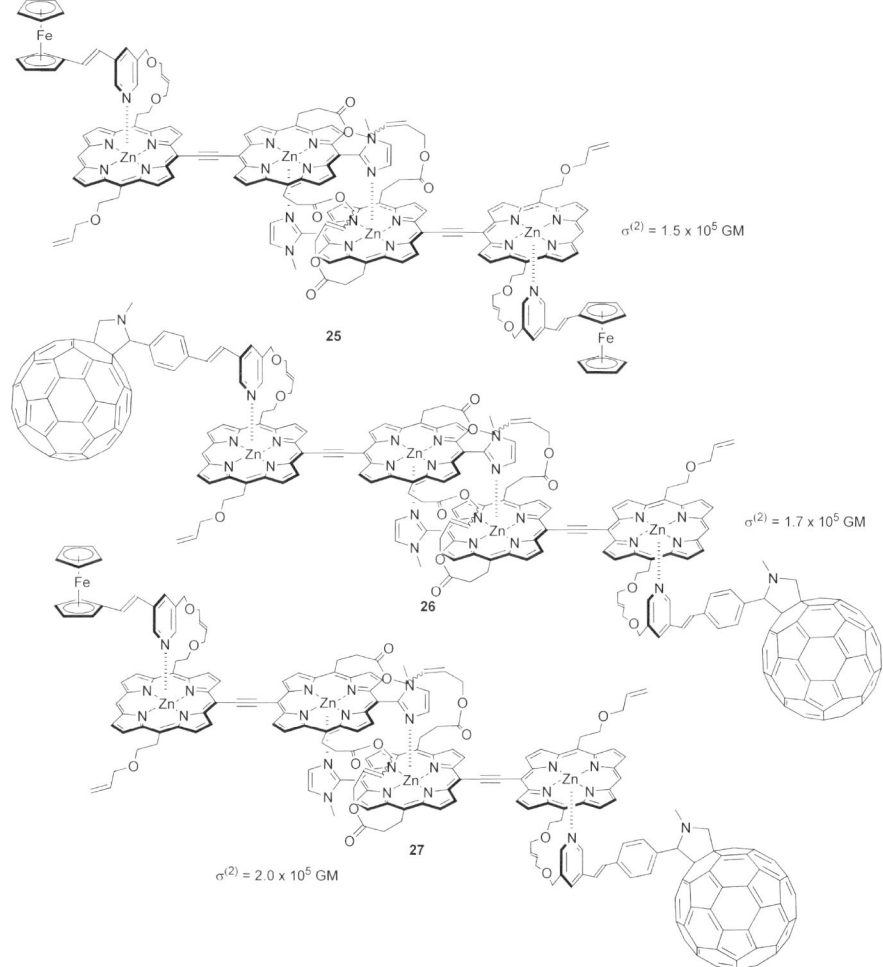

Fig. 16 Self-assembled porphyrin arrays **25–27** with electron donor and acceptor [41]

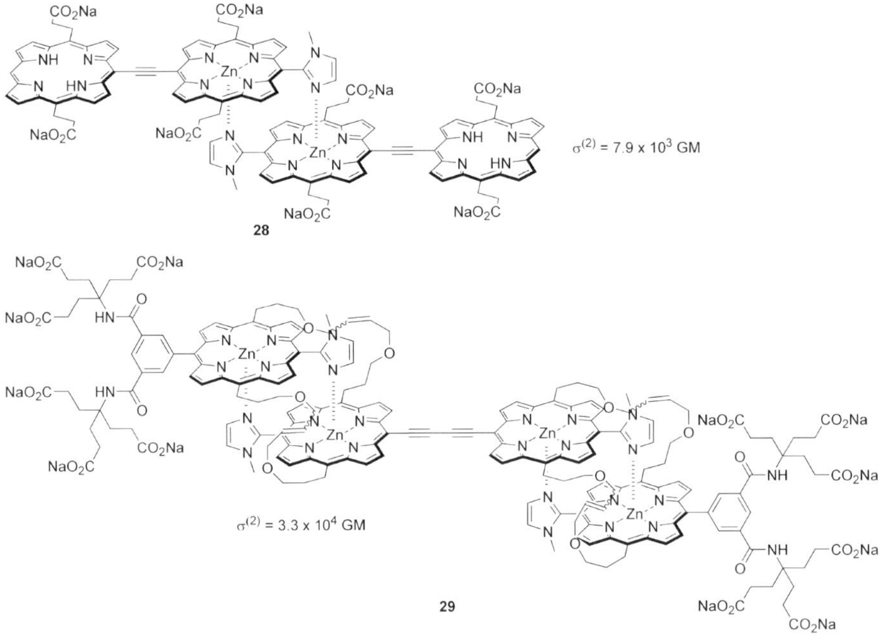

Fig. 17 Water-soluble porphyrin tetramers **28** and **29** [42, 43]

photodynamic therapy because they may allow deeper cancer treatment with a high spatial resolution without damaging healthy cells.

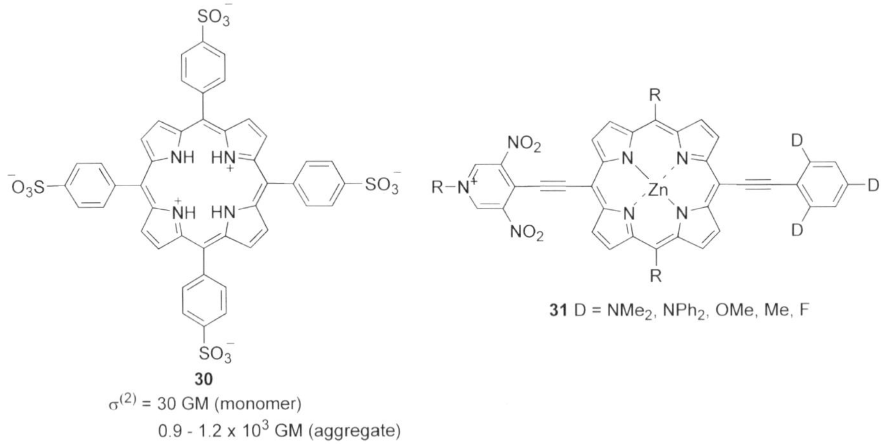

Fig. 18 Molecular structure of tetrakis(4-sulfonatophenyl)porphyrin diacid **30** and push-pull porphyrins **31** [44, 45]

Aggregation of porphyrins has been reported to enhance the TPA cross section. Collini et al. have reported 30-fold enhancement of the TPA cross section (about 10^3 GM) of tetrakis(4-sulfonatophenyl)porphyrin diacid **30** by *J*-type aggregation in water, compared to the corresponding monomer (Fig. 18) [44]. Ray et al. have theoretically predicted the enhancement of TPA cross section ($5.0 \times 10^3 - 9.0 \times 10^3$ GM) by *J*-aggregation of push-pull type porphyrins **31** (ZINDO/CV/SCRF) [45].

3
Artificial Photosynthesis

3.1
Dye-Sensitized Solar Cells and Thin-Film Organic Photovoltaic Cells

Photochemical energy conversion is one of the most promising renewable energy resources. Great efforts have been made to create an organic photochemical energy conversion system due to its low cost and low energy consumption for the large-area, in place of current inorganic photovoltaic cells using silicon semiconductors [46, 47]. This section describes a photoenergy conversion system using supramolecules of porphyrins.

Photoenergy conversion efficiency is the most important indicator of solar cells and is evaluated by energy conversion yield (η) described in Eq. 1;

$$\eta = J_{sc} V_{oc} \text{FF}/I_0 , \tag{1}$$

where J_{sc} is short-circuit photocurrent density, V_{oc} open-circuit photovoltage, FF fill factor, which is described in Eq. 2, and I_0 light intensity.

$$\text{FF} = J_{max} V_{max}/J_{sc} V_{oc} , \tag{2}$$

where J_{max} and V_{max} are the photocurrent density and photovoltage for the maximum power output. Another important indicator is the incident photon-to-current conversion efficiency (IPCE), which is measured as a function of the incident photon current density and is estimated following Eq. 3;

$$\text{IPCE}(\%) = (1240 j_{ph}/\lambda I_0) \times 100 , \tag{3}$$

where j_{ph} is the photocurrent density (A/cm^2), λ monochromatic light, I_0 light intensity (W/cm^2).

Before going into detail, the most promising organic solar cells, dye-sensitized solar cells (DSSC) and thin-film organic solar cells using porphyrins are mentioned briefly. DSSC have attracted much attention over the past decade because of their low production cost and relatively high performance. In TiO$_2$-based dye-sensitized nanocrystalline solar cells, efficiencies up to 11% have been obtained using Ru dyes [48]. However, the limited availability of these dyes together with their undesirable environmental problems has led to the use of cheaper and safer organic-based dyes. Porphyrins have been extensively studied as good candidates for the dyes for DSSC [49]. Schmidt-Mende et al. have reported DSSC using porphyrins **32a** and **32b** instead of Ru dyes and their cell efficiency was 5.2 and 7.1%, respectively (Fig. 19). They revealed that the nature of the carboxylic acid linker to the porphyrin, a combination of a conjugated ethenyl or diethenyl linker in the β-pyrrolic position and a carboxylic binding, had a significant influence on the light-harvesting and photovoltaic properties of the device [50–52].

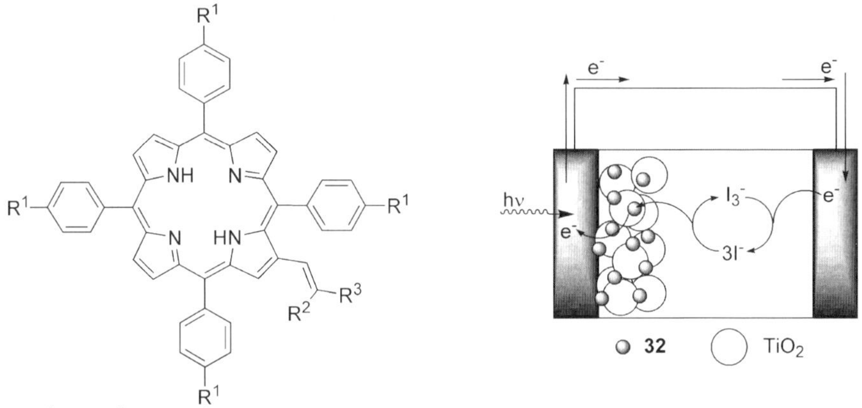

32a (R^1 = H, R^2 = CN, R^3 = CO_2H): η = 5.2%
32b (R^1 = CH_3, R^2 = H, R^3 = CH=C($CO_2H)_2$): η = 7.1%

Fig. 19 Molecular structure and cell efficiencies of porphyrin sensitizers **32**

Thin-film organic photovoltaic cells are also promising candidates for renewable and alternative sources of electrical energy, and therefore increased efforts have been put into the development of solar cells based on small molecules [53] and conjugated polymers [54]. As described in the following section in detail, porphyrins and Pcs tend to form face-to-face aggregates through π–π stacking interaction. Therefore, they are ideal for

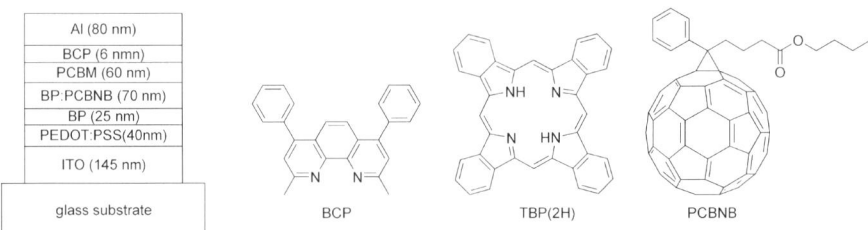

Fig. 20 Structure of p–i–n organic photovoltaic cell

electronic conduction and have been extensively used for plastic organic photovoltaic cells [47, 55–58]. To construct devices using small conducting molecules like porphyrins and Pcs, vacuum deposition is one of the most commonly used procedures [53, 59–61]. For example, Shao et al. have recently developed a heterojunction photovoltaic cell by vacuum deposition using PtOEP and C_{60}, ITO/PEDOT/PtOEP/C_{60}/BCP/Al. The fill factor was 0.57 and power-conversion efficiency was 2.1% [62]. Very recently, Sato et al. have reported p-i-n organic photovoltaics based on solution-processed benzoporphyrin, ITO/PEDOT:PSS/TBP/TBP:fullerene/fullerene/BCP/Al, by spin coating of a soluble precursor of TBP followed by thermal conversion of the precursor to TBP as a film (Fig. 20). The power conversion efficiency was 3.0% [63]. Further details relating to the solution-processed electronic device, organic field-effect transistors (OFETs), and using TBP are presented in the next section.

3.2
Donor–Acceptor Systems by π–π Interactions

Inspired by nature, many kinds of covalently linked D–A compounds including porphyrins and Pcs have been prepared to investigate the mechanism of electron or energy transfer systems [64]. Recently, supramolecules containing porphyrins, Pcs, and their oligomers associated by weak and exchangeable hydrogen bonding, coordinate bonding, π–π interactions, and so on, have been paid a lot of attention, because, compared to covalently linked systems, the supramolecule system can be easily associated by self-assembly of compounds. These supramolecules have been reported for the study of energy and electron transfer in the light-harvesting antenna and the photosynthetic reaction centers [65–68], construction of photovoltaic systems [69, 70], and so on. Since the structures of supramolecules of porphyrins with fullerenes and single-wall carbon nanotubes (SWNTs) are described in detail elsewhere in this volume (see Komatsu N, 2008, in this volume), we will concentrate here on recent topics concerning molecular devices as organic photovoltaic cells or photocurrent-generating systems using supramolecules of porphyrins with acceptors.

Fullerenes have attracted a great deal of attention as good electron acceptors because of their wide-spread π conjugation and their small reorganization energy. Because of the weak π–π donor–acceptor interactions, porphyrins and fullerenes have been characterized to explore CT interactions and photoinduced electron-transfer processes. The structures of cocrystals have been investigated by X-ray crystallographic and theoretical approaches [71–78]. Naturally assembling cocrystallates of fullerenes and tetraphenylporphyrins (TPPs) show unusually short porphyrin/fullerene contacts (2.7–3.0 Å) compared with typical π–π interactions (3.0–3.5 Å) [72] (see Komatsu N, 2008, in this volume).

These supramolecular interactions have been applied to photoenergy conversion systems assembled on nanoparticles of Au, SnO$_2$, and TiO$_2$ [79–92]. Supramolecular complexes of porphyrins and fullerenes are self-assembled to larger clusters by lypophylic interaction, and then associated on a nanos-

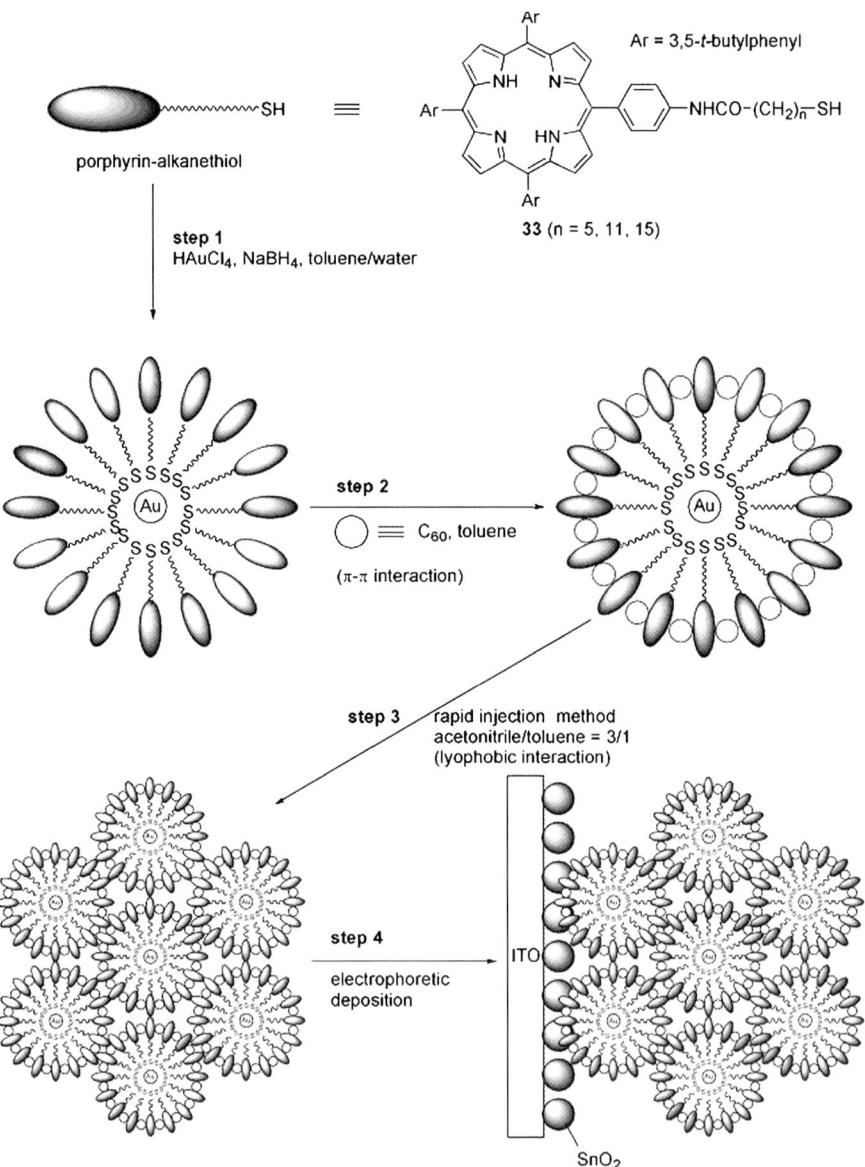

Scheme 1 Bottom-up self-organization of porphyrin **33** and fullerene using gold nanoparticles on a nanostructured SnO$_2$ electrode [91]

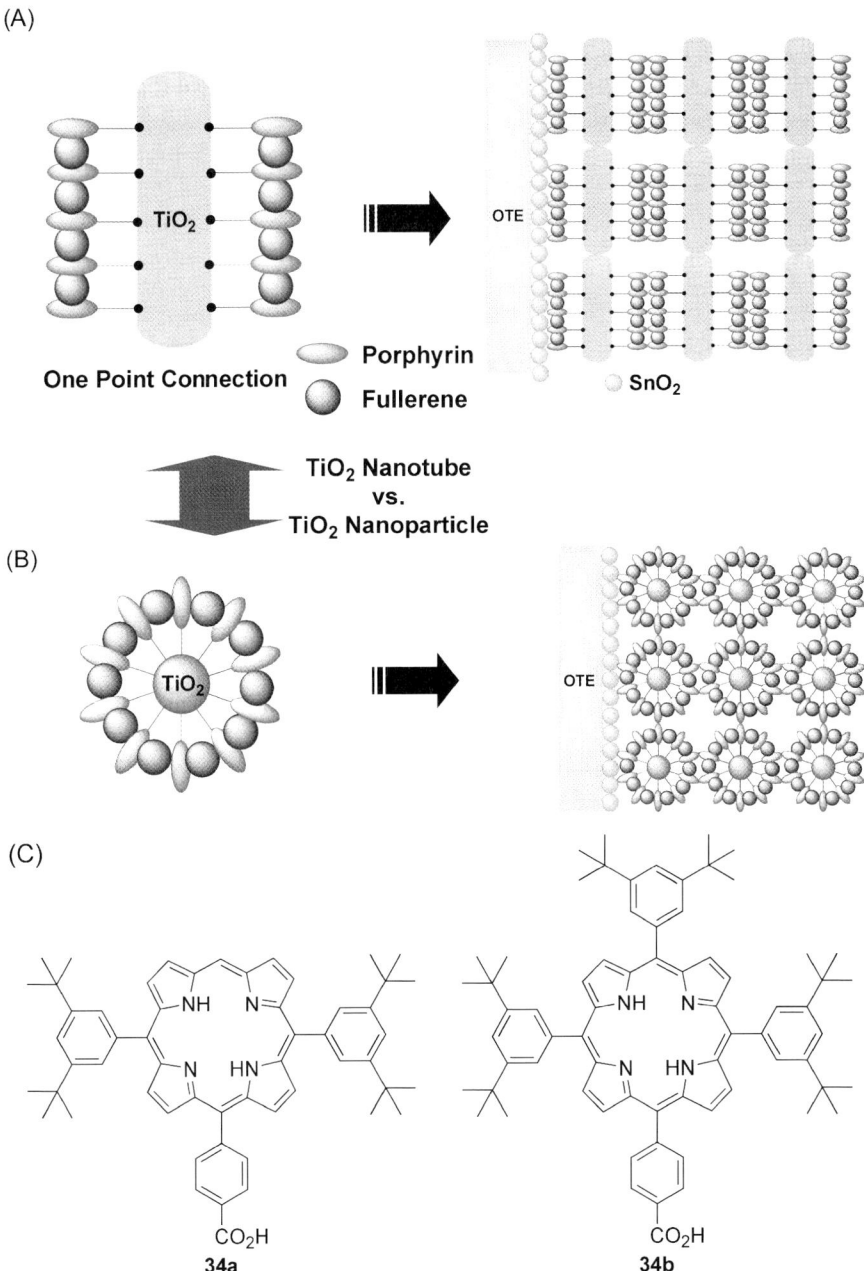

Fig. 21 Diagrammatic summary of the organization between porphyrin **34** and fullerene moieties using TiO$_2$ nanotubes (**A**), nanoparticles (**B**) and molecular structures of **34a** and **34b** (**C**) (reprinted with permission from [93]. © (2007) WILEY-VCH)

tructured SnO$_2$ electrode using an electrophoretic deposition technique (Scheme 1). The best IPCE value of the system, ITO/SnO$_2$/(**33**+C$_{60}$)$_m$/NaI+I$_2$/Pt, was 54%, when the chain length of methylene was 15. The overall power conversion efficiency (η) at an input power (I_o) of 3.4 mW cm^{-2} was 0.61% [79]. In 2007, Hasobe et al. succeeded in the organization of porphyrin–fullerene architecture with TiO$_2$ nanotubes. Using porphyrin **34a**, the maximum IPCE value of 60% was obtained, which was 1.5-times larger than values obtained on TiO$_2$ nanoparticles (Fig. 21) [93]. The IPCE value with **34a** was better than that with **34b**. The results indicated the importance of substrate morphology in promoting electron transport within the mesoscopic semiconductor film. Kang et al. reported the supramolecular assembly of T(3,5-dimethoxy)PP **35a** and fullerene on ITO/SnO$_2$ (Fig. 22) [90]. The modified electrode [ITO/SnO$_2$/(**35a**+C$_{60}$)] showed an IPCE value of 59% at 425 nm, 0.05 V (vs. SCE), in the presence of 3I$^-$/I$_3^-$. When T(3,4,5-trimethoxy)PP **35b** or T(2,6-dimethoxy)PP **35c** were used instead of porphyrin **35a**, IPCE values were 10% and 5.7%, respectively. From UV-Vis absorption spectra, the structure of the J-aggregate was observed only for **35a**/C$_{60}$. The packing structure of **35a**/C$_{60}$ is shown in Fig. 22. The porphyrin and C$_{60}$ molecules define an alternating layered structure where the closest porphyrin moieties (center-to-center distance, Rcc = 14.2 Å) are arranged in a 1D chain with a dihedral angle of 66°, while the closest C$_{60}$ moieties (Rcc = 10.2 Å) are arranged in a 2D sheet.

Hasobe et al. reported on photochemical solar cells using protonated porphyrin-SWNT supramolecules on nanostructured SnO$_2$ electrodes (see

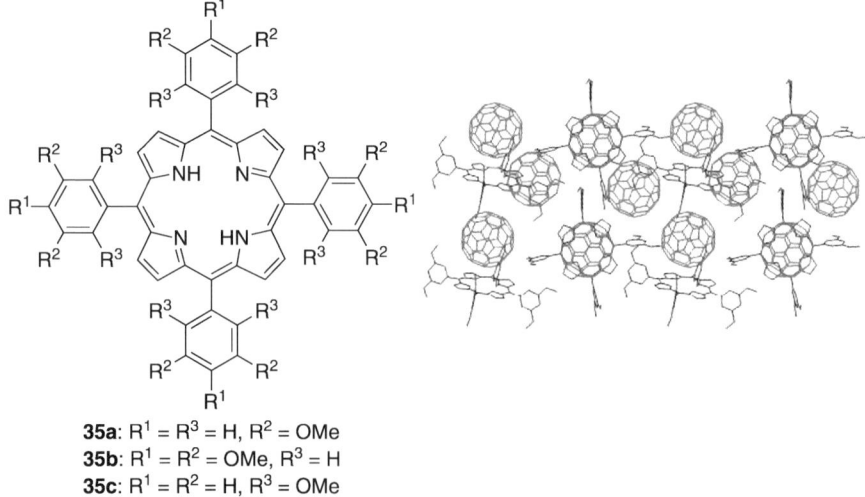

35a: R^1 = R^3 = H, R^2 = OMe
35b: R^1 = R^2 = OMe, R^3 = H
35c: R^1 = R^2 = H, R^3 = OMe

Fig. 22 Molecular structure of porphyrin **35** and molecular packing of **35a**·2C$_{60}$·toluene (reprinted with permission from [90]. © (2006) WILEY-VCH)

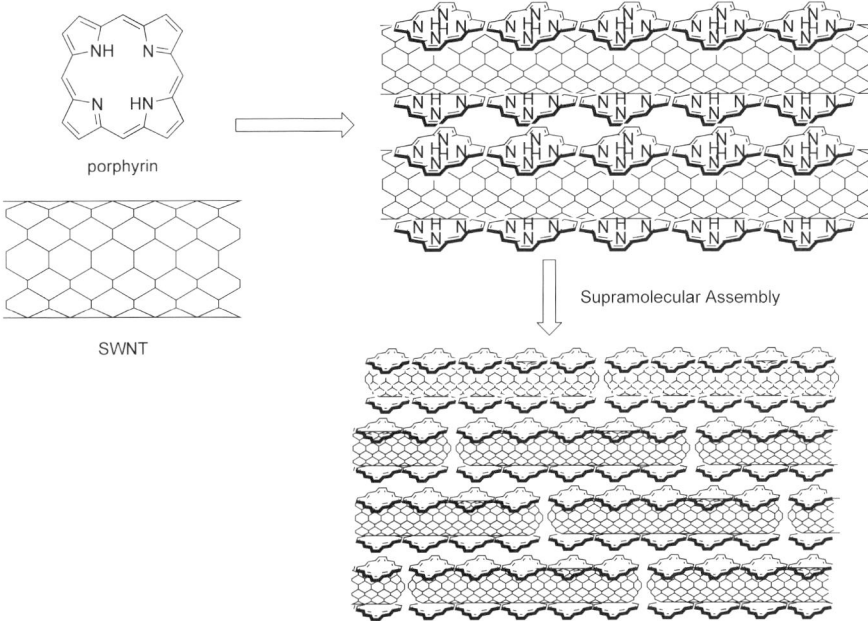

Scheme 2 Self assembly of protonated porphyrins (see Fig. 10 in Komatsu N, 2008, in this volume) and SWNTs by $\pi-\pi$ interaction [95]

Figure 10 from Komatsu N, 2008, in this volume, and Scheme 2) [94, 95]. The mixture of protonated porphyrins and SWNTs were deposited by the electrophoresis technique on nanostructured SnO_2 electrodes and the porphyrins and SWNTs show ordered structure by $\pi-\pi$ stacking. The IPCE of the system was 13% at 0.2 V bias (vs. SCE).

3.3
Photoconversion Systems by Coordination, Hydrogen Bonding and Electrostatic Interactions

Supramolecules of zinc-imidazolyl ligation of zinc imidazolylporphyrin have been applied to photoconversion systems. Self-assembled monolayers (SAM) of imidazole-substituted porphyrins **36** and **37** form, in a supramolecular fashion, a chain structure leading to a significant increase of light absorption in the visible light region and therefore photocurrents (Fig. 23) [96]. Introduction of a porphyrin-bearing electron acceptor onto a SAM increased the photocurrent values [97]. Recently, a systematic series of ferrocene/porphyrin redox cascade architectures were assembled through slipped-cofacial porphyrin dimer **38** on an ITO electrode and the quantum yield in anodic photocurrent generation was 40%, which was the highest value among the reported values on ITO electrodes (Fig. 24) [98].

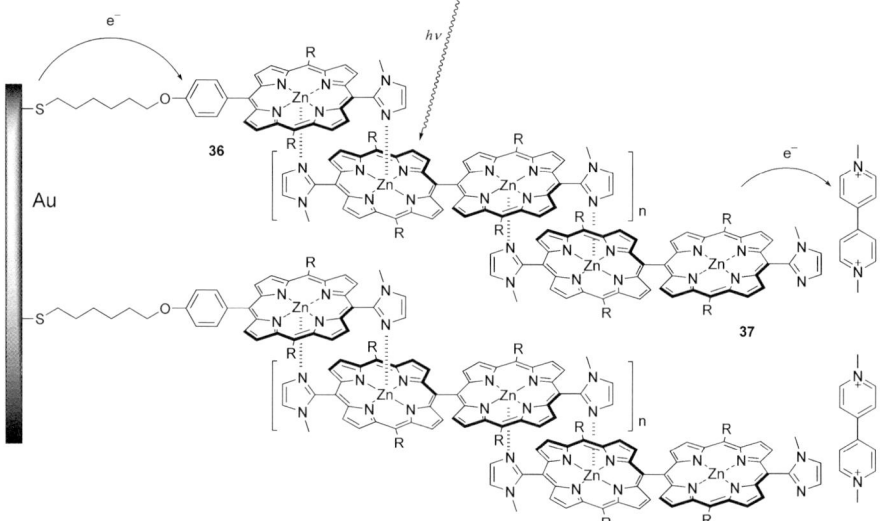

Fig. 23 Coordination assembly of antenna porphyrins **36** and **37** on a gold surface [96]

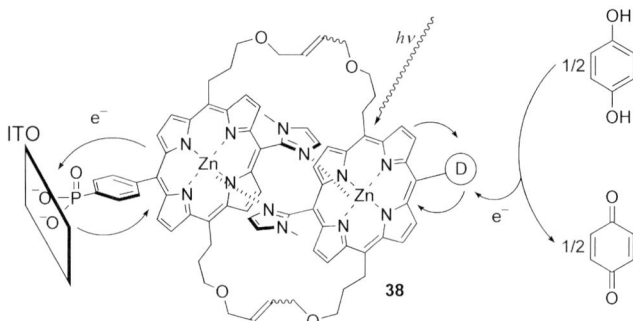

Fig. 24 Schematic illustration of the structure of donor|porphyrin **38**|ITO (D = Ph, Me_8Fc, Fc, $PhMe_8Fc$, PhFc, and C_2Fc) [98]

Drain et al. reported linear porphyrin arrays self-assembled by hydrogen bonding or metal-ion coordination into lipid bilayer membranes [99]. In Fig. 25, linear porphyrin tapes by hydrogen bonding between 5,15-bisdiacetamidopyridyl porphyrins **39** and 5,15-diuracil porphyrins **40** are shown. The size of the porphyrin assembly can self-adjust to the thickness of the bilayer and 85 nA of photocurrent was observed.

Fig. 25 Schematic illustration of one of four self-assembling porphyrin **39** systems self-organized into bilayers (reprinted with permission from [99]. © (2002) The National Academy of Sciences)

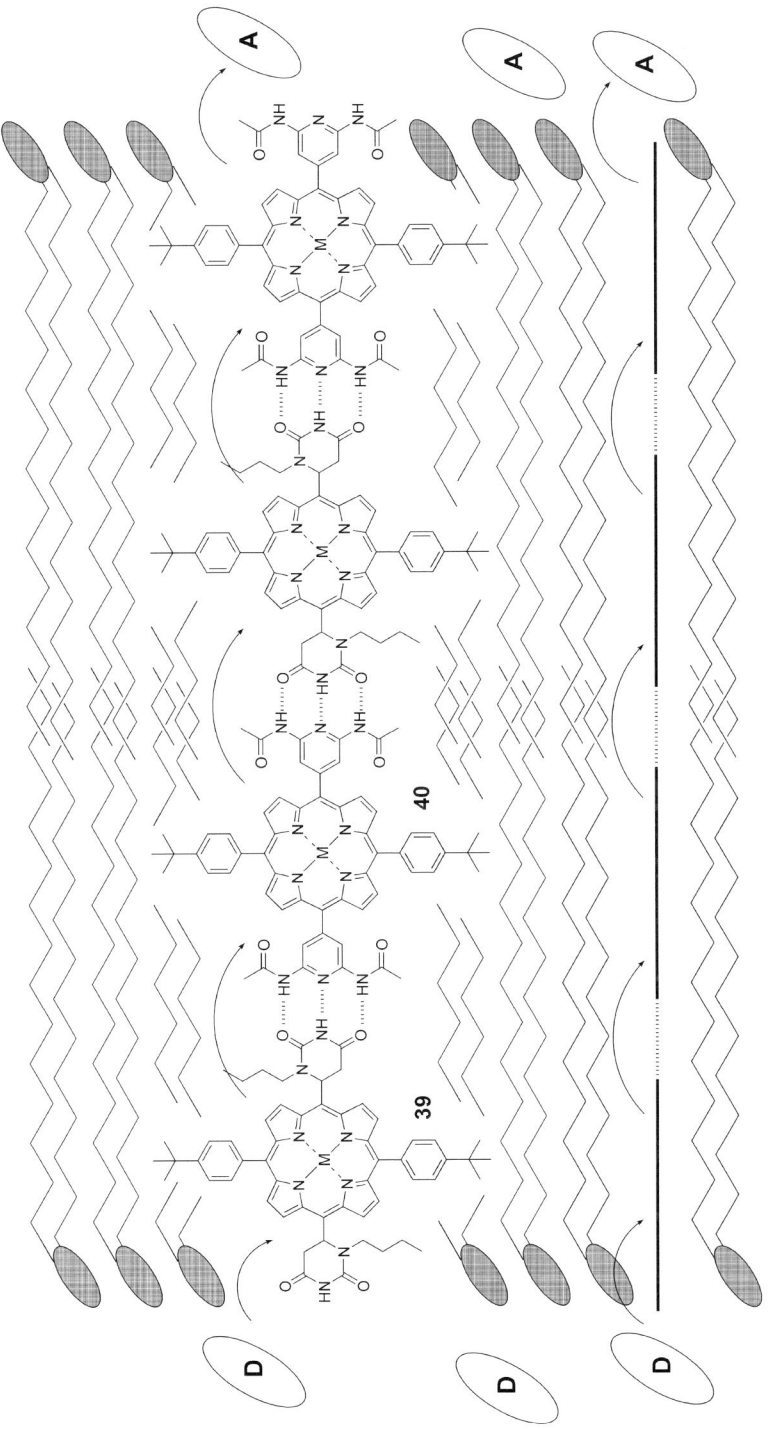

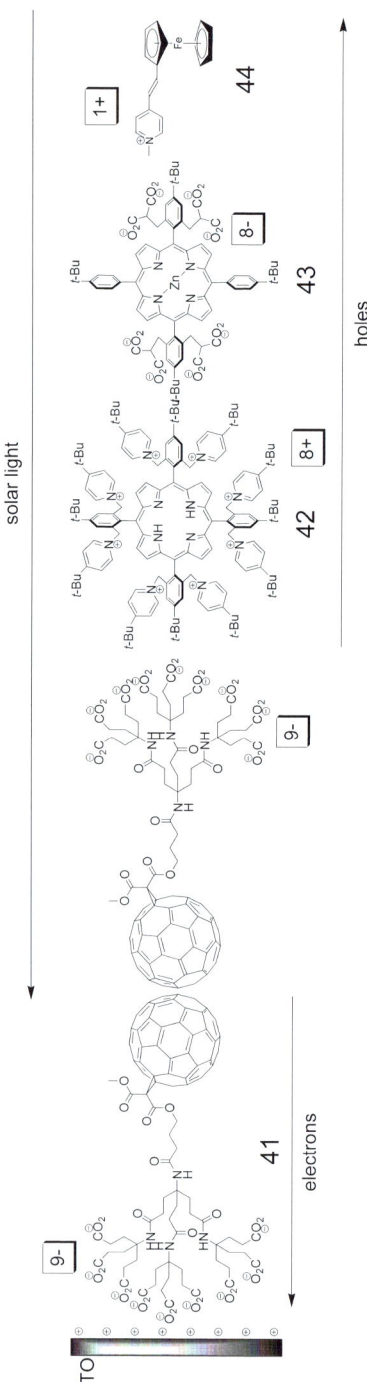

Fig. 26 Electrostatic assembly of fullerene **41**, porphyrins **42**, **43**, and ferrocene **44** [100]

Supramolecules Based on Porphyrins

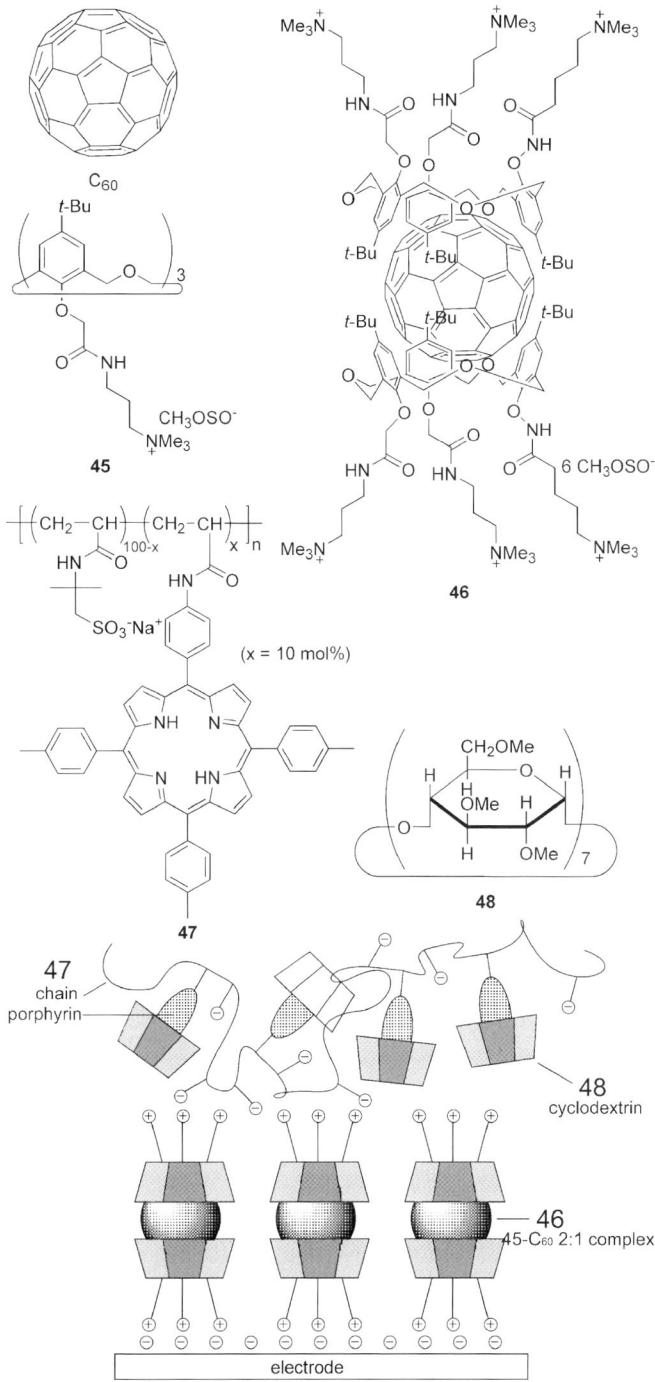

Fig. 27 Schematic illustration of self-assembled multilayers on an ITO electrode [105]

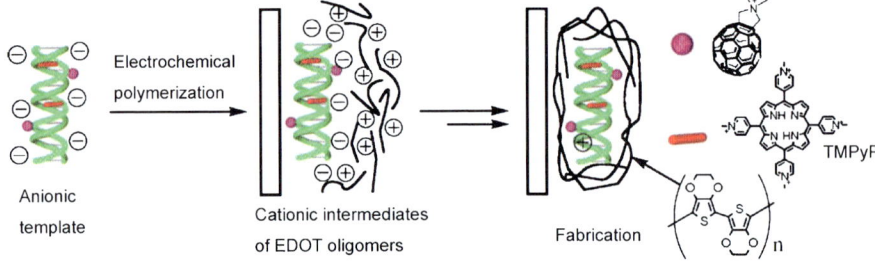

Fig. 28 Schematic illustration of the monocationic fullerene/tetracationicporphyrin/DNA-poly(EDOT) film (reprinted with permission from [107]. © (2005) Elsevier)

Guldi et al. reported the electrostatic and van der Waals assembly of four components, fullerene, free-base porphyrin, zinc porphyrin, and ferrocene, creating redox gradients on ITO-electrodes (Fig. 26) [100]. Electrostatically driven deposition of oppositely charged components increases the flexibility in replacing with individual building blocks, **41**, **42**, **43**, and **44**. The IPCE value of the system shown in Fig. 26 was 1.6%. They also reported on a solar-energy conversion system using SWNT as a building block. SWNT was covered with positively charged pyrene, and then the pyrene interacted with negatively charged porphyrins, ZnP^{8-} (see Scheme 1 from Komatsu N, 2008, in this volume) [101–103]. The ITO electrodes covered with SWNT/pyrene/ZnP^{8-} showed a IPCE value of 8.5% with a –200 mV bias.

Ikeda et al. and Konishi et al. reported on a supramolecular photocurrent generation system in combination with electrostatic and van der Waals interactions. In a C_{60}-porphyrin bilayer prepared by electrostatic alternate adsorption (Fig. 27), the quantum yield of photocurrent generation is increased when self-aggregation of porphyrins is suppressed by host–guest interaction of cyclodextrin and porphyrin [104–106]. Cationic fullerene and tetracationic porphyrin bound on a DNA scaffold by electrostatic interactions was fabricated by conjugate polymer (Fig. 28). The quantum yield of photocurrent generation was 3.8% [107].

4
Electronic Applications

4.1
General Aspects of Organic Field-Effect Transistors (OFETs)

Porphyrins and Pcs tend to form face-to-face aggregates through π–π stacking interactions. Since they accept or donate electrons easily through their large π-electron frameworks, they are suitable for electronic conduction. They have been extensively used as semiconductors in electronic devices such

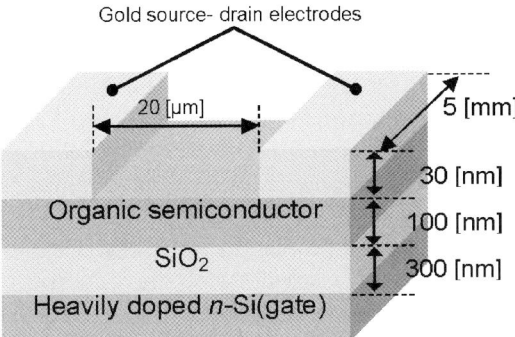

Fig. 29 Illustration of top contact OFET device

as organic light-emitting diodes (OLED), OFETs, and solar cells. During the last decade, research on OFETs has achieved remarkable progress [65, 108–111]. An OFET is illustrated in Fig. 29. The devices consist of deposited conductions (gate, source, and drain electrodes), an insulator (SiO$_2$ or plastics) and organic semiconductors. The OFET works as a switch, with on/off states, when a bias is applied on the gate electrode, the gate is biased negatively or positively to induce hole or electron transport in organic semiconductor layers, respectively. Holes or electrons are transported by the p-channel or n-channel, respectively. In order to apply OFET for driving circuits in display applications, they need to exhibit high carrier output, good switching speed, and high contrast between the on and off states. They are related to several important parameters, namely, carrier mobility (cm^2/Vs), threshold voltage (V_{th}), and on/off current ratio. V_{th} is controlled by subtleties between the organic semiconductor–insulator interfaces that are not well understood. Mo-

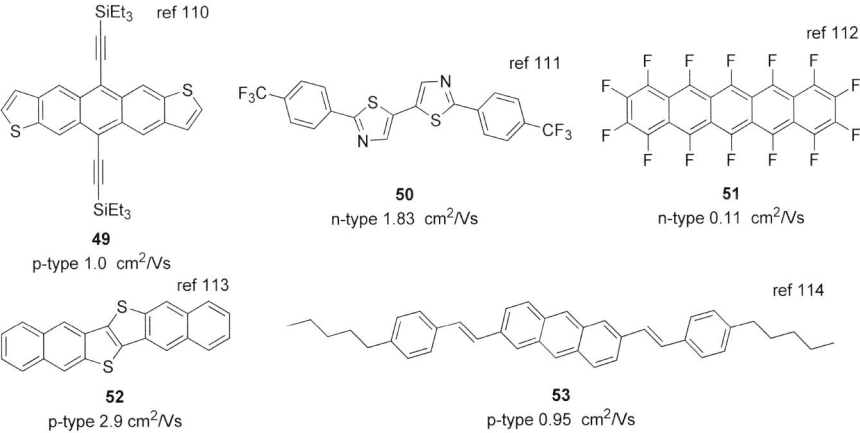

Fig. 30 Recently developed semiconductors for OFET [112–116]

bility is the most important parameter for evaluation of OFET, and the on/off ratio is the second most important one.

Recently, new organic semiconductors have been synthesized to improve carrier mobility, sensitivity, and stability. Some of them are listed in Fig. 30 along with their mobilities [112–116]. The carrier mobilities of organic semiconductors have reached the range of 0.1 to 3 cm^2/Vs, which rivals amorphous silicon (a-Si) devices.

On the other hand, Pcs are promising active materials for OFETs due to their stability, and have been studied widely for a long time. Among various metal complexes CuPc and NiPc show the best mobilities, but they are as low as 0.02 cm^2/Vs [117], values that are too low to be used instead of a-Si. Much effort has been put into improving OFET performance based on Pcs. A sandwich-type thin-film device consisting of two kinds of Pc metal complex displays a mobility of 0.11 cm^2/Vs [118]. The mobility of OFET based on single crystal CuPc is 1 cm^2/Vs, which is the highest reported value so far for Pc-based OFETs [119].

Fine control of crystal structures in the thin film is crucial to improve the performance of OFETs. Nevertheless, extensive efforts have been put into controlling Pcs and related compounds using an LB technique or discotic liquid crystals, and the mobilities are usually in the range of 10^{-3} to 10^{-4} cm^2/Vs [120]. A recent report on OFETs based on rare-earth metal triple-decker complexes of amphiphilic tris(Pcs) shows remarkably high mobilities of 0.24–0.60 cm^2/Vs, as shown in Fig. 31 [121].

Fluorinated copper phthaocyanine (F$_{16}$CuPc) is one of the few molecules that exhibit air-stable n-channel semiconducting behavior. The mobility of the OFET is 0.03 cm^2/Vs when the thin film is fabricated on SiO$_2$ by vapor deposition [122]. The structure of the F$_{16}$CuPc film on SiO$_2$ depends on the thickness of the film, which can be controlled by the deposition rate and

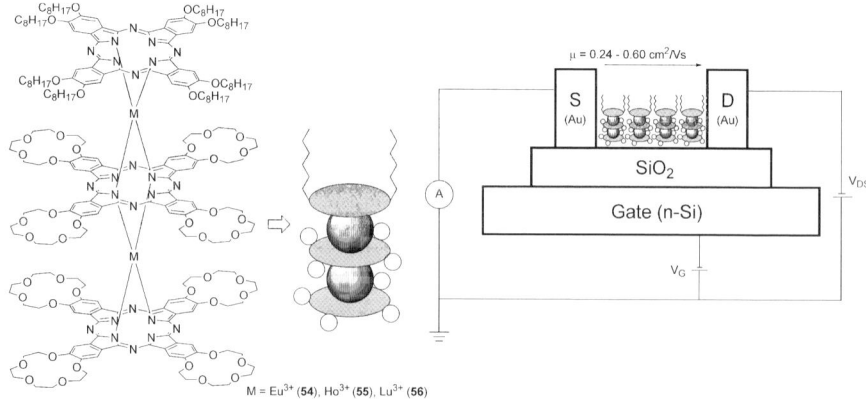

Fig. 31 Structure of triple-decker Pcs and the device [121]

temperature [123]. This suggests that the mobility of FET based on Pcs can be improved by proper choice of fabrication. High-performance air-stable *n*-type OFETs based on single crystalline submicro- and nanometer ribbons of $F_{16}CuPc$ are reported, whose mobility is $0.2\,cm^2/Vs$ [124]. Such single crystalline submicro- and nanometer ribbons can be in situ grown by a physical vapor transport technique along the surface of Si/SiO_2 substrates during fabrication.

4.2 OFETs by Solution-Processed Tetrabenzoporphyrins (TBPs)

Recently, a very useful method for fabrication of TBPs for OFETs has been developed. TBPs are a new semiconductor for OFETs, compared to Pcs. TBPs and Pcs are pigments (insoluble in organic solvents) and it is not easy to get pure samples. Ito et al. have found that heating a porphyrin (CP) fused with bicyclo[2.2.2]octadiene (BCOD) gives TBP in quantitative yield (Scheme 3) [125]. As CPs are soluble in organic solvents, they are purified by column chromatography.

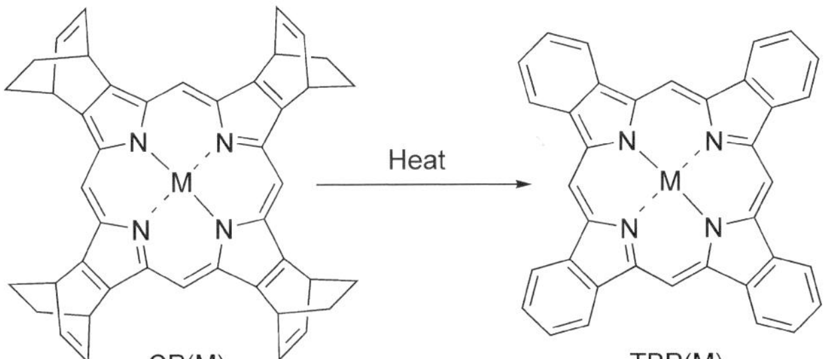

CP(M) TBP(M)

Soluble, Amorphous Thin Film Insoluble, Polycrystalline Thin Film Semiconductors

M = 2H, Zn, Ni, Cu

Scheme 3 Conversion of soluble precursors into insoluble semiconductors

Spin coating of purified CP(2H) followed by heating at 170–200 °C gives an insoluble crystalline semiconductor thin film of TBP(2H). Spun-cast films of the precursor exhibit amorphous, insulating behavior upon a thermal annealing either in vacuum or under N_2, and the amorphous films are converted into polycrystalline films of TBP(2H) with crystal sizes exceeding 1 μm [126]. Observed mobility of the devices exceeds $10^{-2}\,cm^2/Vs$ with appropriate process, device structure, and on/off current ratios exceeding 10^5 (Fig. 32) [126].

OFETs using metal porphyrins such as TBP(Cu) [126] or TBP(Ni) [127] are also fabricated by a solution process using the corresponding soluble precursors, and they offer better performances in OFETs than that of TBP(2H) itself. XRD patterns for drop-cast CP(Ni) and TBP(Ni) thin film are shown in

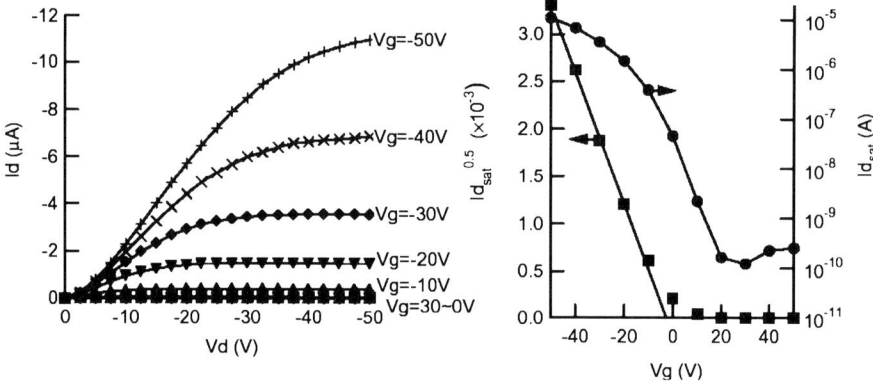

Fig. 32 Transfer characteristics of a TBP OFET on SiO_2 [126]

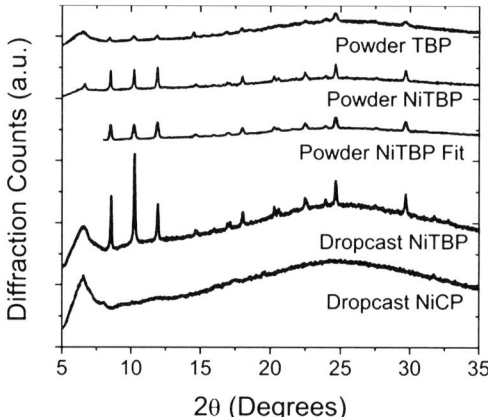

Fig. 33 XRD spectra for NiTBP and CP(Ni) powder and thin films

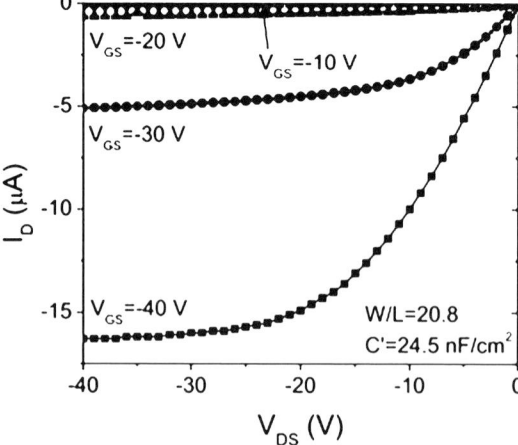

Fig. 34 Output characteristics of a NiTBP OFET [128]

Fig. 33. The pattern for drop-cast CP(Ni) displays no measurable peaks. Upon heating, numerous diffraction peaks indicative of the formation of crystal planes appear. OFETs output characteristics are shown in Fig. 34, the mobility of this transistor being of the order of 0.1 and 0.2 cm^2/Vs, which is the highest value among solution-processed OFETs using porphyrins or Pcs as semiconductors. OFETs based on TBP(Cu) exhibit a similar performance with the mobility of 0.1 cm^2/Vs. A polarized optical micrograph of a spun-cast TBP(Cu) thin film is shown in Fig. 35 and the polycrystalline nature of the TBP(Cu) thin films is displayed. The electrodes are 20 μm wide, indicating

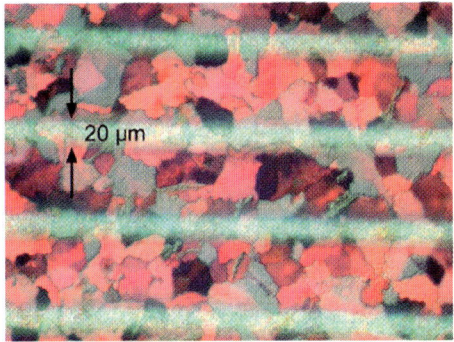

Fig. 35 Polarized optical micrograph of a continuous, spun-cast TBP(Cu) thin-film on thermally oxidized c-Si. The electrodes in the figure are 20 μm wide [127]

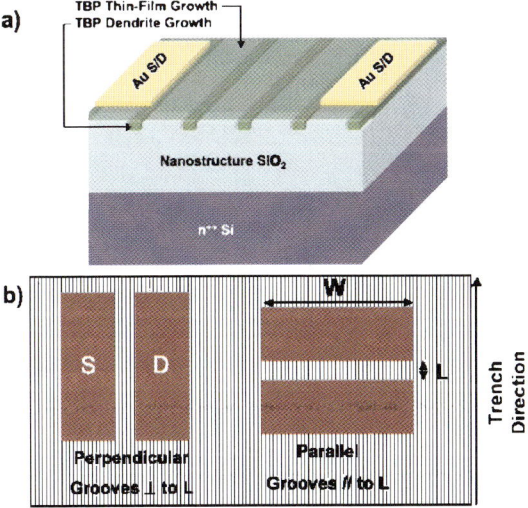

Fig. 36 a Schematic cross section of nanostructured TBP OFETs. **b** Top view schematic of nanostructured OFET structures (reprinted with permission from [129]. © (2007) American Institute of Physics)

that TBP(Cu) forms domains of approximately the same size. By comparison, TBP(Ni) forms crystals approaching 1 mm [128], and TBP(2H) approximately 2 μm in diameter [126].

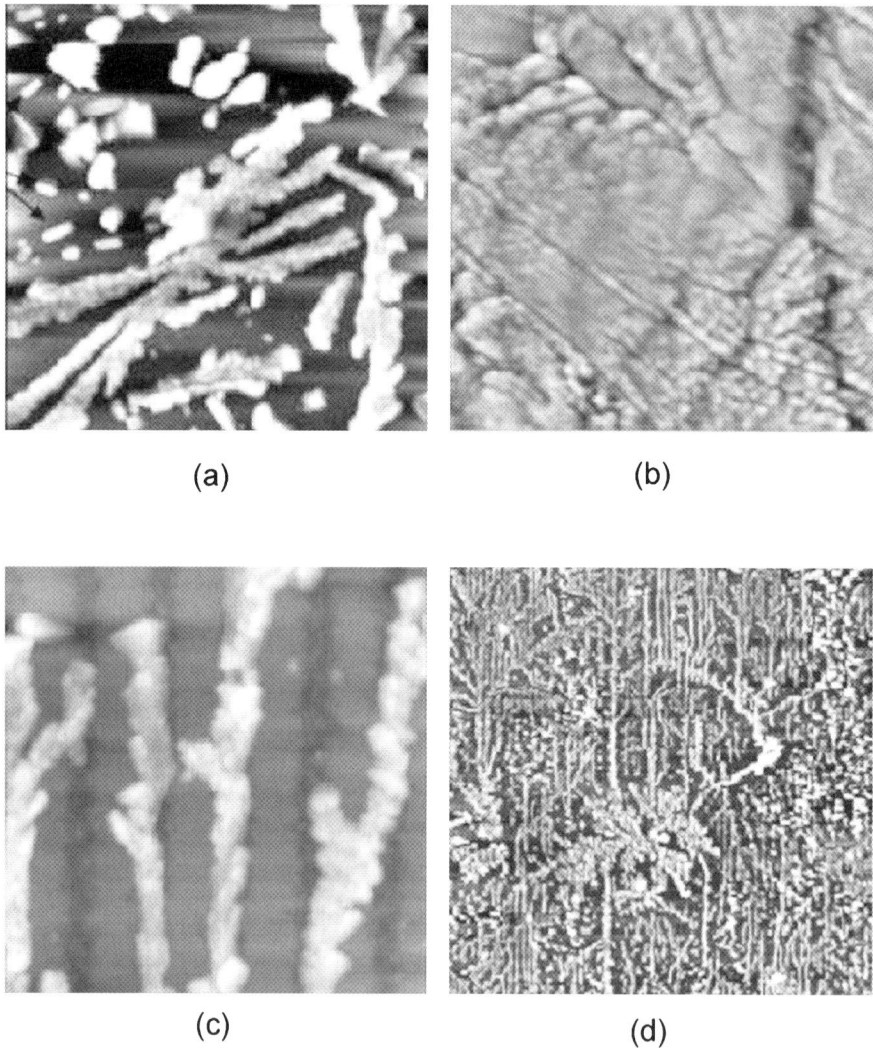

Fig. 37 AFM height micrographs of **a** fractal aggregation of sparse, solution-processed TBP; **b** aligned domain formation in more dense TBP; **c** a close view of TBP aggregation within trenches; and **d** a wider-scale view of TBP aggregation wire formation within trenches. The axes dimensions are **a** 3 μm, **b** 2 μm, **c** 2.2 μm, and **d** 25.2 μm. Samples in (**c**) and (**d**) have a trench periodicity of 450 nm and a trench depth of 10 nm (reprinted with permission from [129]. © (2007) American Institute of Physics)

Solution-processed OFETs are fabricated using a precursor form (CPs) of TBPs deposited on a thermal SiO_2 gate insulator patterned with nanometer-scale trenches (Fig. 36). Thermal conversion of CP to TBP is enhanced by ordered TBP aggregation in the prepatterned trenches. OFETs with channels parallel to trench direction growth are found to have field-effect mobility approaching one order of magnitude greater than transistors fabricated with the channel oriented perpendicular to dendrimer growth [129].

Fractal and micrometer-scale ordered nanorod aggregation are both observed in precursor-route TBP thin film deposited on unpatented bare SiO_2 as shown in Figs. 37a and 37b. Thinner regions of the same film, with sparser surface coverage density, display fractal aggregation. For thin films with higher surface coverage density, nanorod-type aggregation is predominant. Typical TBP rod-shaped aggregates demonstrate widths of 60 nm and lengths of 300 nm. AFM height micrographs of TBP thin film are shown in Figs. 37c and 37d. Preferential alignment along the trench direction is observed upon thermal conversion of CP to TBP.

5
Concluding Remarks

Herein, we describe recent developments in the use of porphyrins and phthalocyanines in the field of optical and electronic materials, the properties of which are mainly controlled by molecular self-assembly. Photovoltaic cells are currently of broad interest as potential low-cost approaches to solar energy conversion. Large-area electronic devices and solution-processed organic semiconductors based on porphyrins, phthalocyanines, and other molecules could have potentially a huge cost advantage over Si-based devices if conversion efficiency and durability can be improved to the level of Si-solar cells. We are now approaching this goal as shown herein, where supramolecules based on porphyrins and phthalocyanines may play crucial roles.

References

1. Special issue on supramolecular chemistry and self-assembly (2002) Science 295:2395
2. Special issue on supramolecular approaches to organic electronics and nanotechnology (2006) Adv Mater 18:1227
3. van Nostrum CF, Nolte RJM (1996) Chem Commun, p 2385
4. Elemans JAAW, van Hameren R, Nolte RJM, Rowan AE (2006) Adv Mater 18:1251
5. Saito T, Sisk W, Kobayashi T, Suzuki S, Iwayanagi T (1993) J Phys Chem 97:8026
6. Tamiaki H (1996) Coord Chem Rev 148:183
7. Hu X, Ritz T, Damjanoviæ A, Schuten K (1997) J Phys Chem B 101:3854
8. Satake A, Kobuke Y (2007) Org Biomol Chem 5:1679

9. Nakamura Y, Aratani N, Osuka A (2007) Chem Soc Rev 36:831
10. Kelly RF, Goldsmith RH, Wasielewski MR (2007) J Am Chem Soc 129:6384
11. Sanders JKM (1996) In: Atwood JL, Davies JED, MacNicol DD, Vögtle F, Lehn J-M (eds) Comprehensive Supramolecular Chemistry, vol 9. Pergamon, Oxford, p 131
12. Sanders JKM (2000) In: Kadish KM, Smith KM, Roger G (eds) The Porphyrin Handbook, vol 3. Academic Press, San Diego, p 347
13. Chambron JC, Herz V, Sauvage JP (2000) In: Kadish KM, Smith KM, Roger G (eds) The Porphyrin Handbook, vol 6. Academic Press, San Diego, p 1
14. Satake A, Kobuke Y (2005) Tetrahedron 61:13
15. Balaban TS (2005) Acc Chem Res 38:612
16. Iengo E, Zangrando E, Alessio E (2006) Acc Chem Res 39:841
17. Rucareanu S, Schuwey A, Gossauer A (2006) J Am Chem Soc 128:3396
18. Hoffmann M, Wilson CJ, Odell B, Anderson HL (2007) Angew Chem Int Ed 46:3122
19. Kobuke Y, Miyaji H (1994) J Am Chem Soc 116:4111
20. Kobuke Y, Ogawa K (2003) Bull Chem Soc Jpn 76:689
21. Kameyama K, Morisue M, Satake A, Kobuke Y (2005) Angew Chem Int Ed 44:4763
22. Maeda C, Shinokubo H, Osuka A (2007) Org Lett 9:2493
23. Kamada T, Aratani N, Ikeda T, Shibata N, Higuchi Y, Wakamiya A, Yamaguchi S, Kim KS, Yoon ZS, Kim D, Osuka A (2006) J Am Chem Soc 128:7670
24. Matano Y, Matsumoto K, Terasaka Y, Hotta H, Araki Y, Ito O, Shiro M, Sasamori T, Tokitoh N, Imahori H (2007) Chem Eur J 13:891
25. Sessler JL, Jayawickramarajah J, Gouloumis A, Torres T, Guldi DM, Maldonado S, Stevenson KJ (2005) Chem Commun, p 1892
26. Shirakawa M, Kawano S, Fujita N, Sada K, Shinkai S (2003) J Org Chem 68:5037
27. Shirakawa M, Fujita N, Shinkai S (2003) J Am Chem Soc 125:9902
28. Shirakawa M, Fujita N, Shinkai S (2005) J Am Chem Soc 127:4164
29. O'Flaherty SM, Hold SV, Cook MJ, Torres T, Chen Y, Hanack M, Blau WJ (2003) Adv Mater 15:19
30. Chou J-H, Kosal ME, Nalwa HS, Rakow NA, Suslick KS (2000) In: Kadish KM, Smith KM, Roger G (eds) The Porphyrin Handbook, vol 6. Academic Press, San Diego, p 73
31. Drobizhev M, Stepanenko Y, Dzenis Y, Karotki A, Rebane A, Taylor PN, Anderson HL (2004) J Am Chem Soc 126:15352
32. Drobizhev M, Stepanenko Y, Dzenis Y, Karotki A, Rebane A, Taylor PN, Anderson HL (2005) J Phys Chem B 109:7223
33. Karotki A, Drobizhev M, Dzenis Y, Taylor PN, Anderson HL, Rebane A (2004) Phys Chem Chem Phys 6:7
34. Ahn TK, Kim KS, Kim DY, Noh SB, Aratani N, Ikeda C, Osuka A, Kim D (2006) J Am Chem Soc 128:1700
35. Kim DY, Ahn TK, Kwon JH, Kim D, Ikeue T, Aratani N, Osuka A, Shigeiwa M, Maeda S (2005) J Phys Chem A 109:2996
36. Inokuma Y, Ono N, Uno H, Kim DY, Noh SB, Kim D, Osuka A (2005) Chem Commun, p 3782
37. Screen TEO, Thorne JRG, Denning RG, Bucknell DG, Anderson HL (2002) J Am Chem Soc 124:9712
38. Drobizhev M, Stepanenko Y, Rebane A, Wilson CJ, Screen TEO, Anderson HL (2006) J Am Chem Soc 128:12432
39. Ogawa K, Ohashi A, Kobuke Y, Kamada K, Ohta K (2003) J Am Chem Soc 125:13356
40. Ogawa K, Ohashi A, Kobuke Y, Kamada K, Ohta K (2005) J Phys Chem B 109:22003
41. Tanihara J, Ogawa K, Kobuke Y (2006) J Photochem Photobiol A 178:140

42. Ogawa K, Hasegawa H, Inaba Y, Kobuke Y, Inouye H, Kanemitsu Y, Kohno E, Hirano T, Ogura S, Okura I (2006) J Med Chem 49:2276
43. Dy JT, Ogawa K, Satake A, Ishizumi A, Kobuke Y (2007) Chem Eur J 13:3491
44. Collini E, Ferrante C, Bozio R (2005) J Phys Chem B 109:2
45. Ray PC, Sainudeen Z (2006) J Phys Chem A 110:12342
46. Grätzel M (2004) J Photochem Photobiol A 164:3
47. Peumans P, Yakimov A, Forrest SR (2003) J Appl Phys 93:3693
48. O'Regan B, Grätzel M (1991) Nature 353:737
49. Campbell WM, Burrell AK, Officer DL, Jolley KW (2004) Coord Chem Rev 248:1363
50. Schmidt-Mende L, Campbell WM, Wang Q, Jolley KW, Officer DL, Nazeeruddin MK, Grätzel M (2005) ChemPhysChem 6:1253
51. Wang Q, Campbell WM, Bonfantani EE, Jolley KW, Officer DL, Walsh PJ, Gordon K, Humphry-Baker R, Nazeeruddin MK, Grätzel M (2005) J Phys Chem B 109:15397
52. Campbell WM, Jolley KW, Wagner P, Wagner K, Walsh PJ, Gordon KC, Schmidt-Mende L, Nazeeruddin MK, Wang Q, Grätzel M, Officer DL (2007) J Phys Chem C 111:11760
53. Tang CW (1986) Appl Phys Lett 48:183
54. Yu G, Gao J, Hummelen JC, Wudl F, Heeger AJ (1995) Science 270:1789
55. Brabec CJ, Sariciftci NS, Hummelen JC (2001) Adv Funct Mater 11:15
56. Nelson J (2002) Curr Opin Solid State Mater Sci 6:87
57. Goetzberger A, Hebling C, Schock HW (2003) Mater Sci Eng R 40:1
58. Forrest SR (2004) Nature 428:911
59. Yamashita K (1982) Chem Lett 1085
60. Yamashita K, Harima Y, Kubota H, Suzuki H (1987) Bull Chem Soc Jpn 60:803
61. Wienke J, Schaafsma TJ, Goossens A (1999) J Phys Chem B 103:2702
62. Shao Y, Yang Y (2005) Adv Mater 17:2841
63. Sato Y, Niinomi T, Hashiguchi M, Matsuo Y, Nakamura E (2007) SPIE Proc 6656
64. Gust D, Moore TA (2000) In: Kadish KM, Smith KM, Roger G (eds) The Porphyrin Handbook, vol 8. Academic Press, San Diego, p 153
65. Wojaczyński J, Latos-Grażyński L (2000) Coord Chem Rev 204:113
66. El-Khouly ME, Ito O, Smith PM, D'Souza F (2004) J Photochem Photobiol C 5:79
67. Mateo-Alonso A, Sooambar C, Maurizio P (2006) D R Chimie 9:944
68. D'Souza F, Ito O (2005) Coord Chem Rev 249:1410
69. Konishi T, Ikeda A, Shinkai S (2005) Tetrahedron 61:4881
70. Bonifazi D, Enger O, Diederich F (2007) Chem Soc Rev 36:390
71. Sun Y, Drovetskaya T, Bolskar RD, Bau R, Boyd PDW, Reed CA (1997) J Org Chem 62:3642
72. Boyd PDW, Hodgson MC, Rickard CEF, Oliver AG, Chaker L, Brothers PJ, Bolskar RD, Tham FS, Reed CA (1999) J Am Chem Soc 121:10487
73. Sun D, Tham FS, Reed CA, Chaker L, Burgess M, Boyd PDW (2000) J Am Chem Soc 122:10704
74. Zheng J-Y, Tashiro K, Hirabayashi Y, Kinbara K, Saigo K, Aida T, Sakamoto S, Yasmaguchi K (2001) Angew Chem Int Ed 40:1857
75. Sun D, Tham FS, Reed CA, Boyd PDW (2002) Proc Natl Acad Sci USA 99:5088
76. Wang Y-B, Lin Z (2003) J Am Chem Soc 125:6072
77. Ayabe M, Ikeda A, Kubo Y, Takeuchi M, Shinkai S (2002) Angew Chem Int Ed 41:2790
78. Boyd PDW, Reed CA (2005) Acc Chem Res 38:235
79. Hasobe T, Imahori H, Kamat PV, Fukuzumi S (2003) J Am Chem Soc 125:14962
80. Hasobe T, Imahori H, Fukuzumi S, Kamat PV (2003) J Phys Chem B 107:12105

81. Hasobe T, Kamat PV, Absalom MA, Kashiwagi Y, Sly J, Crossley MJ, Hosomizu K, Imahori H, Fukuzumi S (2004) J Phys Chem B 108:12865
82. Hasobe T, Kashiwagi Y, Absalom MA, Sly J, Hosomizu K, Crossley MJ, Imahori H, Kamat PV, Fukuzumi S (2004) Adv Mater 16:975
83. Hasobe T, Imahori H, Kamat PV, Ahn TK, Kim SK, Kim D, Fujimoto A, Hirakawa T, Fukuzumi S (2005) J Am Chem Soc 127:1216
84. Imahori H, Fujimoto A, Kang S, Hotta H, Yoshida K, Umeyama T, Matano Y, Isoda S (2005) Adv Mater 17:1727
85. Hasobe T, Imahori H, Kamat PV, Ahn TK, Kim SK, Kim D, Fujimoto A, Hirakawa T, Fukuzumi S (2005) J Am Chem Soc 127:1216
86. Imahori H, Mitamura K, Shibano Y, Umeyama T, Matano Y, Yoshida K, Isoda S, Araki Y, Ito O (2006) J Phys Chem B 110:11399
87. Hasobe T, Hattori S, Kamat PV, Fukuzumi S (2006) Tetrahedron 62:1937
88. Imahori H, Mitamura K, Umeyama T, Hosomizu K, Matano Y, Yoshida K, Isoda S (2006) Chem Commun, p 406
89. Imahori H, Fujimoto A, Kang S, Hotta H, Yoshida K, Umeyama T, Matano Y, Isoda S (2006) Tetrahedron 62:1955
90. Kang S, Umeyama T, Ueda M, Matano Y, Hotta H, Yoshida K, Isoda S, Shiro M, Imahori H (2006) Adv Mater 18:2549
91. Imahori H (2007) Bull Chem Soc Jpn 80:621
92. Imahori H (2007) J Mater Chem 17:31
93. Hasobe T, Fukuzumi S, Hattori S, Kamat PP (2007) Chem Asian J 2:265
94. Hasobe T, Fukuzumi S, Kamat PV (2006) J Phys Chem B 110:25477
95. Hasobe T, Fukuzumi S, Kamat PV (2005) J Am Chem Soc 127:11884
96. Nomoto A, Kobuke Y (2002) Chem Commun, p 1104
97. Nomoto A, Mitsuoka H, Ozeki H, Kobuke Y (2003) Chem Commun, p 1074
98. Morisue M, Kalita D, Haruta N, Kobuke Y (2007) Chem Commun, p 2348
99. Drain CM (2002) PNAS 99:5178
100. Guldi DM, Zilbermann I, Anderson G, Li A, Balbinot D, Jux N, Hatzimarinaki M, Hirsch A, Prato M (2004) Chem Commun, p 726
101. Guldi D, Rahman GMA, Jux N, Tagmatarchis N, Prato M (2004) Angew Chem Int Ed 43:5526
102. Guldi DM (2005) J Phys Chem B 109:11432
103. Ehli C, Rahman GMA, Jux N, Balbinot D, Guldi DM, Paolucci F, Marcaccio M, Paolucci D, Melle-Franco M, Zerbetto F, Campidelli S, Prato M (2006) J Am Chem Soc 128:11222
104. Ikeda A, Hatano T, Shinkai S, Akiyama T, Yamada S (2001) J Am Chem Soc 123:4855
105. Ikeda A, Hatano T, Konishi T, Kikuvhi J-I, Shinkai S (2003) Tetrahedron 59:3537
106. Konishi T, Ikeda A, Asai M, Hatano T, Shinkai S, Fujitsuka M, Ito O, Tsuchiya Y, Kikuchi J-I (2003) J Phys Chem B 107:11261
107. Bae A-H, Hatano T, Sugiyasu K, Kishida T, Takeuchi M, Shinkai S (2005) Tetrahedron Lett 46:3169
108. Murphy AR, Fréchet JMJ (2007) Chem Rev 107:1066
109. Würthner F, Schmidt R (2006) ChemPhysChem 7:793
110. Loo Y-L (2007) AIChE J 53:1066
111. Takimiya K, Kunugi Y, Otsubo T (2007) Chem Lett 36:578
112. Payne MM, Parkin SR, Anthony JE, Kuo C-C, Jackson TN (2005) J Am Chem Soc 127:4986
113. Ando S, Murakami R, Nishida J, Tada H, Inoue Y, Tokito S, Yamashita Y (2005) J Am Chem Soc 127:14996

114. Sakamoto Y, Suzuki T, Kobayashi M, Gao Y, Fukai Y, Inoue Y, Sato F, Tokito S (2004) J Am Chem Soc 126:8138
115. Yamamoto T, Takimiya K (2007) J Am Chem Soc 129:2224
116. Meng H, Sun F, Goldfiner MB, Gao F, Londono DJ, Marshal WJ, Blackman GS, Dobbs KD, Keys DE (2006) J Am Chem Soc 128:9304
117. Bao Z, Lovinger AJ, Dodabalapur A (1997) Adv Mater 9:42
118. Zhang J, Wang J, Wang H, Yan D (2004) Appl Phys Lett 84:142
119. Zeis R, Siegrist T, Kloc C (2005) Appl Phys Lett 86:022103
120. Xiao K, Liu Y, Huang X, Xu Y, Yu G, Zhu D (2003) J Phys Chem B 107:9226
121. Chen Y, Su W, Bai M, Jianzhuang J, Liu X, Wang L, Wang S (2005) J Am Chem Soc 127:15700
122. Bao Z, Lovinger AJ, Brown J (1998) J Am Chem Soc 120:207
123. de Oteyza DG, Barrena E, Ossó J, Sellner S, Dosch H (2006) J Am Chem Soc 128:15052
124. Tang Q, Li H, Liu Y, Hu W (2006) J Am Chem Soc 126:14634
125. Ito S, Murashima T, Uno H, Ono N (1998) Chem Commun, p 1661
126. Aramaki S, Sakai Y, Ono N (2004) Appl Phys Lett 84:2085
127. Shea PB, Pattison LR, Kawano M, Chen C, Chen J, Petroff P, Matin DC, Yamada H, Ono N, Kanicki J (2007) Synth Metals 157:190
128. Shea PB, Kanicki J, Pattison LR, Petroff P, Kawano M, Yamada H, Ono N (2006) J Appl Phys 100:034502
129. Shea PB, Chen C, Kanicki J, Pattison LR, Petroff P (2007) Appl Phys Lett 90:233107

Heterocyclic Supramolecular Chemistry of Fullerenes and Carbon Nanotubes

Naoki Komatsu

Department of Chemistry, Shiga University of Medical Science, Seta, 520-2192 Otsu, Japan
nkomatsu@belle.shiga-med.ac.jp

1	Introduction	162
1.1	Chemistry of Fullerenes	162
1.2	Chemistry of Carbon Nanotubes	162
1.3	Structures and Terminology of Carbon Nanotubes	163
2	Exohedral Supramolecular Chemistry of Fullerenes and Carbon Nanotubes	167
2.1	Supramolecules of Porphyrins and the Analogues with Fullerenes	167
2.1.1	Porphyrin Monomers and Related Heterocycles	167
2.1.2	Porphyrin Oligomers	169
2.2	Supramolecules of Porphyrins and the Analogues with Carbon Nanotubes	172
2.2.1	Porphyrin Monomers and the Analogues	172
2.2.2	Porphyrin Oligomers and Polymers	178
2.3	Supramolecules of DNA with Carbon Nanotubes	183
2.4	Supramolecules of Proteins with Carbon Nanotubes	185
2.5	Supramolecules of Peptides with Carbon Nanotubes	185
2.6	Supramolecules of Carbohydrates with Fullerenes	186
2.7	Supramolecules of Carbohydrates with Carbon Nanotubes	187
3	Endohedral Supramolecular Chemistry of Fullerenes and Carbon Nanotubes	189
4	Concluding Remarks	190
	References	191

Abstract The chemistry of fullerenes and carbon nanotubes have highly contributed to the progress in the fundamental and applies science of these nanomaterials over the last 15 years. This review focuses on the non-covalent chemistry of fullerenes and carbon nanotubes with nitrogen and/or oxygen containing heterocyclic molecules such as porphyrin, DNA, protein, peptide and carbohydrate. Not only exohedral but also endohedral functionalization is reviewed, because the above guest molecules can interact with both faces of the carbon nanotubes. New terminology is also proposed in the structural and stereochemistry of carbon nanotubes.

Keywords Carbon nanotubes · Complexation · Fullerenes · Non-covalent functionalization · Porphyrin

Abbreviations

CD	Cyclodextrin
CNT	Carbon nanotube
CVD	Chemical vapor deposition
DNA	Deoxyribonucleic acid
DWNT	Double-walled carbon nanotube
HRTEM	High-resolution transmission electron microscopy
MWNT	Multi-walled carbon nanotube
PEG	Poly(ethylene glycol)
RNA	Ribonucleic acid
SWNT	Single-walled carbon nanotube

1
Introduction

1.1
Chemistry of Fullerenes

Fullerenes were discovered by Kroto et al. in 1985 [1] and first synthesized in macroscopic quantity by Krätschmer et al. in 1990 [2]. The chemistry of fullerenes began with synthesis, separation, characterization and determination of the fundamental properties. Then, covalent and non-covalent functionalizations were explored extensively. Owing to the good solubility of fullerenes in some organic solvents, pure forms of C_{60} and C_{70} are easily obtained in large amounts by separation of an as-produced fullerene mixture [3–7]. In addition, the methodology in organic chemistry can be directly applied to the chemistry of fullerenes in terms of manipulation, functionalization, separation and analyses. In this sense, fullerenes are not like carbon materials such as diamond, graphite and CNTs, but like organic molecules such as aromatic hydrocarbons. In fact, a series of organic reactions realized the chemical synthesis of C_{60} [8, 9] and encapsulation of molecular hydrogen in C_{60} through opening and closing of the cage [10]. The similarity in availability and behavior of fullerenes to organic compounds has brought about great progress in the science and technology of fullerenes.

This review focuses on the heterocyclic molecules such as porphyrins and the analogues (Sect. 2.1), and carbohydrates (Sect. 2.6) as host molecules for fullerenes. Therefore, references are provided here for other excellent receptors for fullerenes such as calixarenes [11–19], cyclotriveratrylene [15, 20–22] and carbon nanorings [23–28].

1.2
Chemistry of Carbon Nanotubes

In contrast to the fullerenes mentioned above, CNTs are not soluble in any solvents and structurally pure ones are not available, making the chemistry of CNTs different in principle from that of fullerenes. Their insolubility and structural heterogeneity have greatly hampered the progress of nanoscience and nanotechnology of CNTs.

In order to circumvent the solubility problem, CNTs were covalently or non-covalently functionalized, enabling easy handling, improved reactivity and various spectral analyses of SWNTs in solution phase [29–32]. Once they are functionalized, however, it is very difficult to recover the original form by removing the added functionalities. Moreover, the covalent functionalization unavoidably leads to a disruption of the π-conjugate sidewall structure, resulting in drastic changes in their optical and electrical properties. Although the supramolecular method preserves the integrity of the CNT structures, the solubilizing agent used may act as an impurity in the subsequent application. Many drawbacks still remain to be solved in the solubilization of CNTs.

As for the structural homogeneity of CNTs, on the other hand, it is well known that physical properties of SWNTs are closely correlated with their structures. Therefore, SWNTs with controlled structure are in great demand for electrical and optical applications. SWNTs consisting of a single roll-up index have been separated quite recently by use of DNA as the solubilizing agent and size-exclusion chromatography [33]. However, more time is needed to realize a large-scale supply of structurally controlled SWNTs in pure forms. Therefore, every experiment has been carried out using a mixture of CNTs, making the evaluation and interpretation of the outcome more complicated. Even the fundamental physical properties of CNTs have been determined experimentally as a mixture of many structures. As mentioned above, fullerenes have great advantages in that point. In conclusion, the progress in nanoscience and nanotechnology of CNTs is largely dependent on the availability of bulk quantities of CNTs with limited kinds of structures, preferably a single structure.

This review focuses mainly on non-covalent functionalization of CNTs [34] with heterocyclic molecules including artificial and bio-originated molecules. The functionalization is performed not only on the tube (exohedral), but also in the tube (endohedral). CNTs were found to encapsulate a variety of materials for many purposes, which will be described in Sect. 3.

1.3
Structures and Terminology of Carbon Nanotubes

The structures of CNTs are defined by a variety of parameters such as number of carbon layers, length, diameter and alignment of hexagons. Broadly, CNTs are classified into the following three types according to the number of carbon layers; SWNTs, DWNTs and MWNTs. DWNTs and MWNTs were discovered by Iijima at the negative end of the electrode of the arc discharge for the fullerene synthesis in 1991 [35]. SWNTs were first prepared independently by Iijima et al. [36] and Bethune et al. [37]. However, there is controversy regarding "who should be given the credit for the discovery of carbon nanotubes?" [38–42]. All these three types of CNTs were selectively synthesized by use of different combinations of metal catalysts under different conditions [43–49].

Among the three types of CNTs, SWNTs attract growing interest because of their simple structures and electronic properties [50, 51]. The structure of a SWNT can be prepared by rolling up a graphene sheet into a seamless cylinder. The structure is defined by a roll-up vector C_h given by two unit vectors a_1 and a_2; $C_h = na_1 + ma_2$, where n and m are integers and designated as the roll-up index (n, m) as shown in Fig. 1 [52, 53]. The following three types of SWNTs can be formed; zigzag, armchair and chiral as shown in Fig. 2, depending on the alignment of the hexagonal rings along the tube axis. The (n, m) and C_h have been referred to as "chiral" index (or simply "chirality") and "chiral" vector, respectively, in CNT society. However, the meaning of "chiral" is not always consistent with the original meaning of chiral in chemistry; that is, "the geometric property of a rigid object of being non-superposable on its mirror image" [54]. While chiral SWNTs have non-superposable mirror-image structures; namely, left- and right-handed ones as shown in Fig. 3, zigzag and armchair structures do not have their non-superposable mirror-image. Whether a SWNT is chiral or achiral under the definition in chemistry, the terms of the "chiral" index, "chirality" and "chiral" vector have been used to define its structure in CNT society. As Strano

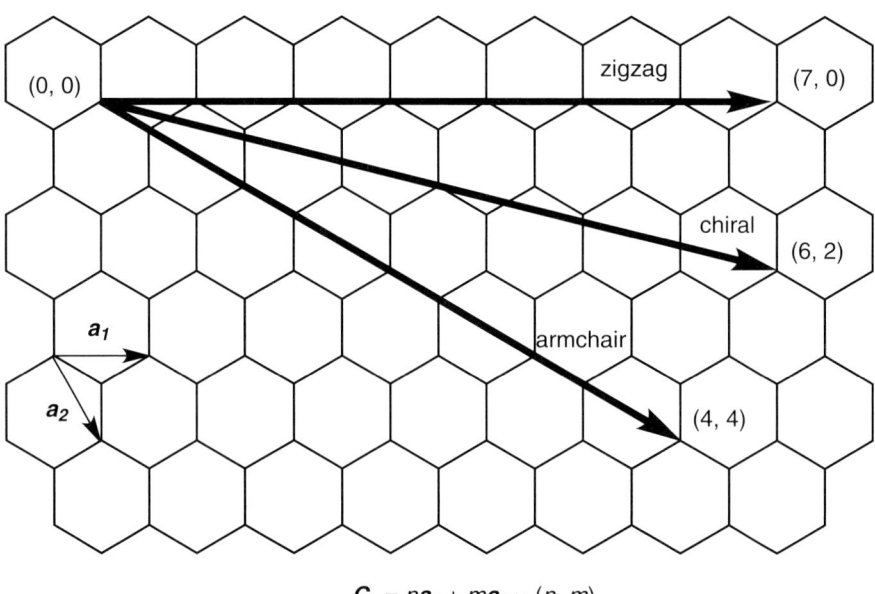

Fig. 1 (n, m) SWNT defined by rolling up the graphene along the roll-up vector C_h [52, 53]

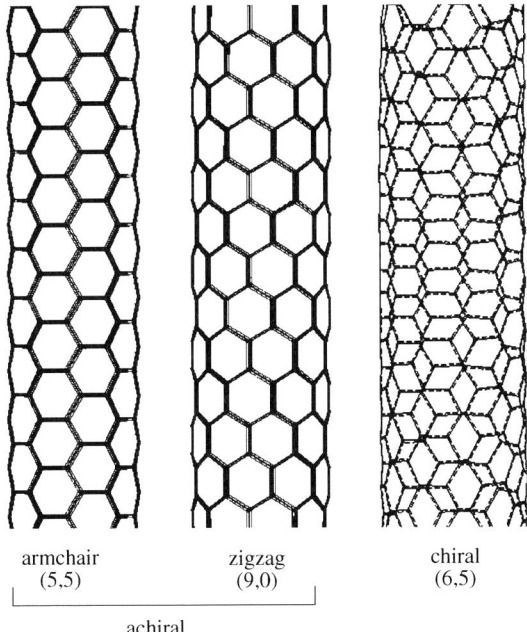

Fig. 2 Armchair, zigzag, chiral structures of SWNTs [52, 53]

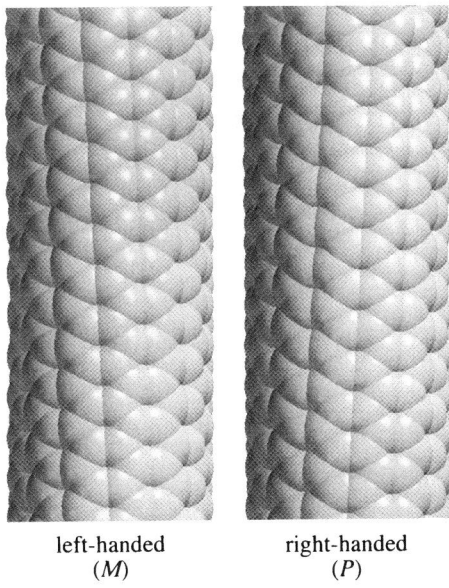

Fig. 3 Chiral (6,5) SWNTs with left- and right-handed structures corresponding to M and P helical structures according to IUPAC nomenclature [54, 57, 58]

pointed out in the article [55], this terminology is very confusing and, hence, a systematic nomenclature is desired to define the structures of CNTs. The terms "roll-up" index and "roll-up" vector are proposed to be used instead of "chiral" index and "chiral" vector, respectively, to indicate (n, m) [56–58]. In this review, the term chiral is used exclusively in the meaning of having the non-superposable property for its mirror image. The "roll-up" index (or simply index) indicates the (n, m) of a CNT and does not include any information about stereochemistry (chirality), which will be discussed below.

Quite recently, optically active SWNTs have been obtained through the optical resolution of their left- and right-handed helical structures by the author's group [56]. For the expression of the stereoisomers of chiral SWNTs, however, a variety of terms have been used so far; *LH* and *RH* [56], *r* and *l* [59], *L* and *R* [60], (n_1, n_2) and (n_2, n_1) [61], and *AL* and *AR* [52]. Since the stereochemistry of SWNTs will be discussed much more than before optically active SWNTs were obtained, a definite nomenclature is required for the stereochemistry of SWNTs [55]. In a recent paper by the author's group [56–58], left- and right-handed helical structures on the basis of the definition of *AL* and *AR* [52] are referred to as *M* and *P*, respectively, according to IUPAC terminology [54]. Every SWNT has three armchair lines (A lines) as indicated by the solid arrows in Fig. 4. These A lines cannot be superposed on its mirror-image in the case of chiral SWNTs, while they can be superposed

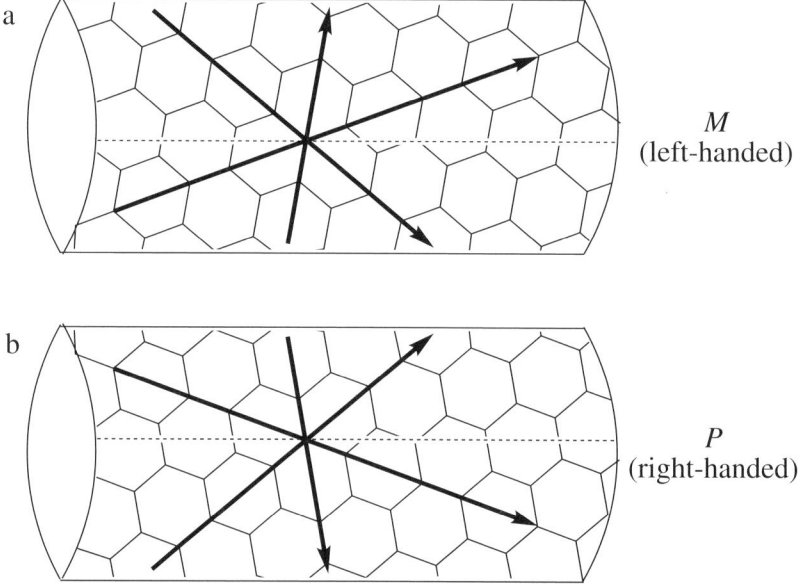

Fig. 4 Definition of *M* (left-handed) and *P* (right-handed) SWNTs. *Three arrows* and *dashed line* indicate armchair lines (A lines) and SWNT axis, respectively [52]

in zigzag and armchair types. When two out of the three A lines are rotated to the left and the third A line to the right, the chiral SWNT is designated as *M* as shown in Fig. 4a. Similarly, the chiral SWNT with two A lines rotated to the right is designated as *P* as shown in Fig. 4b. This review follows this terminology.

2
Exohedral Supramolecular Chemistry of Fullerenes and Carbon Nanotubes

2.1
Supramolecules of Porphyrins and the Analogues with Fullerenes

2.1.1
Porphyrin Monomers and Related Heterocycles

Porphyrins are functional dyes with unique chemical, physical and biological features. Although they are almost flat molecules as shown in Fig. 5, the tetraphenylporphyrin **1**-linked silica stationary phase was found to be effective for chromatographic separation of fullerenes with a curved π-surface in 1993 [62–65]. A close contact (less than 3.0 Å) between the porphyrin plane and curved fullerene surface was observed in the crystal structures of the complex of C_{60}-octakis(dimethylamino)porphyrazine **2** in 1995 [66] and covalently linked fullerene–porphyrin conjugate **3** in 1997 [67] (Fig. 5). The strong interaction was confirmed by many cocrystallites of octaethylporphyrins **4** [68, 69], **1** and its derivatives [70–74], metal tetrahexylporphyrins **5** [75], metal tetrakis(4-pyridyl)porphyrins **6** [76, 77], and dendritic porphyrins **7** [78, 79] with C_{60}, $C_{60}O$, C_{70} $C_{120}O$, methanofullerene and endohedral metallofullerenes (C_{80}) [80–82] (Fig. 5). The nature of the interaction was considered to be electrostatic between an electron negative 6-6 juncture of the fullerene and the electron positive center of the porphyrin as well as π–π interaction between them [82, 83].

Some other nitrogen-containing heterocycles were reported to have affinity to fullerenes. Metal tetraazaannulenes **8** (Fig. 6) encapsulated C_{60} and C_{70} in their crystal structures [84, 85]. Novel types of heterocyclic host molecules designed to capture fullerenes, azacalix[2]arene[2]pyridine **9**, azacalix[4]arene[4]pyridine **10** and highly phenylated triamino-*s*-triazine **11** (Fig. 6), were synthesized, characterized and examined in their complexation behavior with C_{60} and C_{70} [86, 87]. They were found to exhibit very large binding constants; 7.0×10^4 dm^3·mol^{-1} in the complexation of **10** and C_{60}, 1.4×10^5 dm^3·mol^{-1} for **10** and C_{70}, and 0.9–1.8×10^5 dm^3·mol^{-1} for **11** and C_{60}. Azacrown compounds with aryl substituents, **12** and **13** (Fig. 6), formed 1:1 complexes with C_{60} and C_{70}, giving homogeneous LB multilayers with an average bilayer thickness of approximately 47 Å for **12** and 37–38 Å for **13** [88].

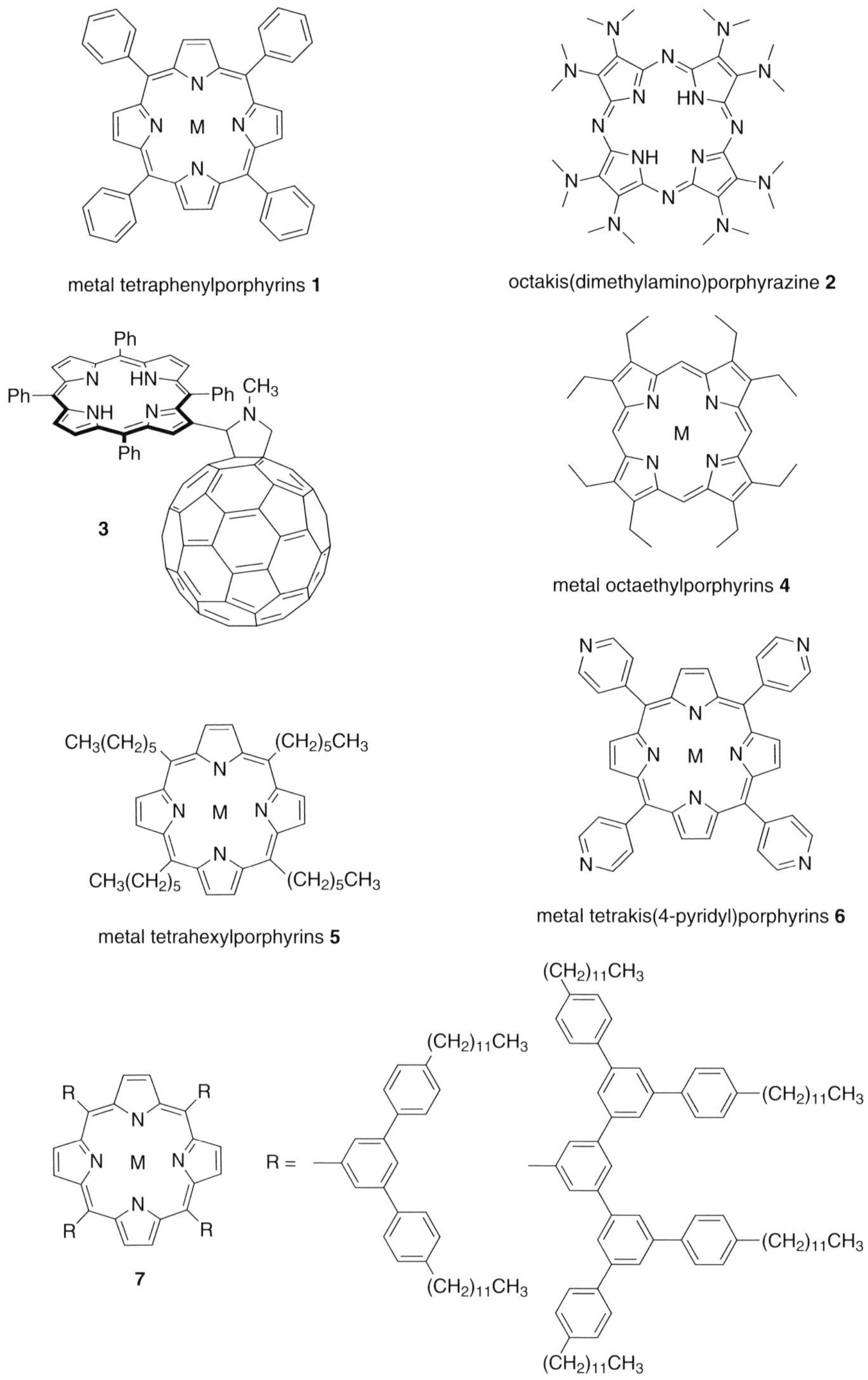

Fig. 5 Structures of porphyrin monomers 1–7 [66–79]

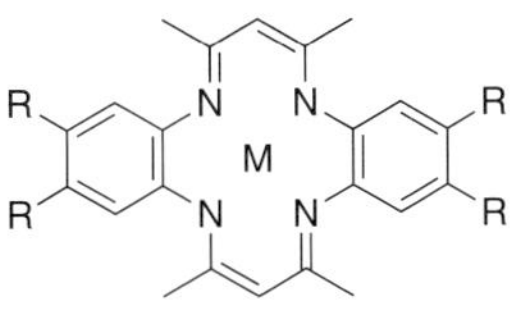

R = H or CH_3, M = Ni, Cu or Zn

metal tetraazaannulene **8**

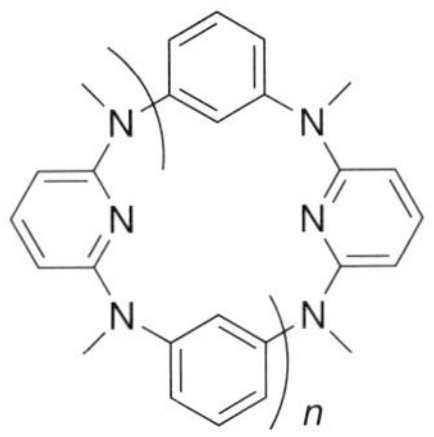

azacalix[*n*]arene[*n*]pyridines
9; *n* = 1
10; *n* = 3

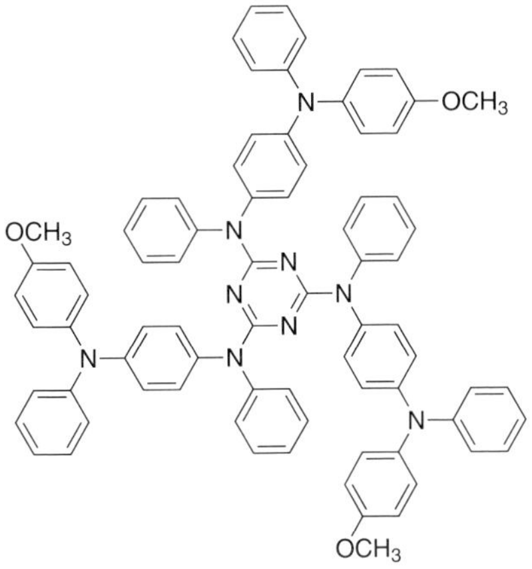

triamino-*s*-triazine **11**

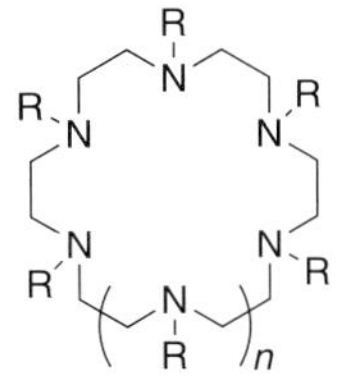

azacrowns

12; R = 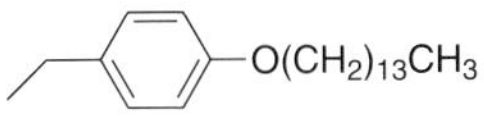 *n* = 1

13; R = 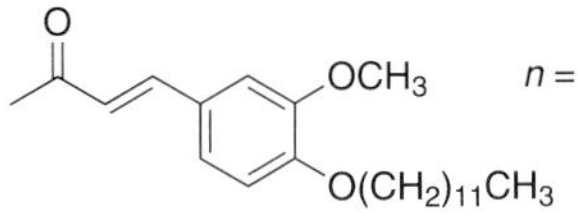 *n* = 3

Fig. 6 Structures of nitrogen-containing heterocycles **8–13** [84–88]

2.1.2 Porphyrin Oligomers

For fullerene encapsulation, host molecules consisting of two or more porphyrin units have been developed, such as cyclic diporphyrins **14** [89–96], gable-type diporphyrins [97, 98] **15** [99], **16** [73], **17** [83, 100], **18** [101, 102] and **19** [103], porphyrin tetramer **20** [104], and porphyrin hexamer **21** [105] (Figs. 7 and 8). They gave very large binding constants and, in some cases, good C_{60}/C_{70} selectivities [82]. A triply fused porphyrin dimer having a similar structure to that of **43** (Fig. 12) is reported to give two-dimensional supramolecular assemblies composed of the dimeric porphyrin and C_{60} [106].

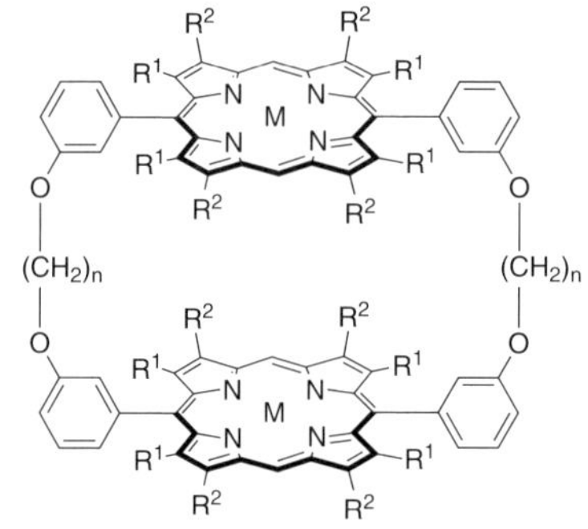

Fig. 7 Structures of porphyrin dimers **14–19** [73, 83, 89–103]

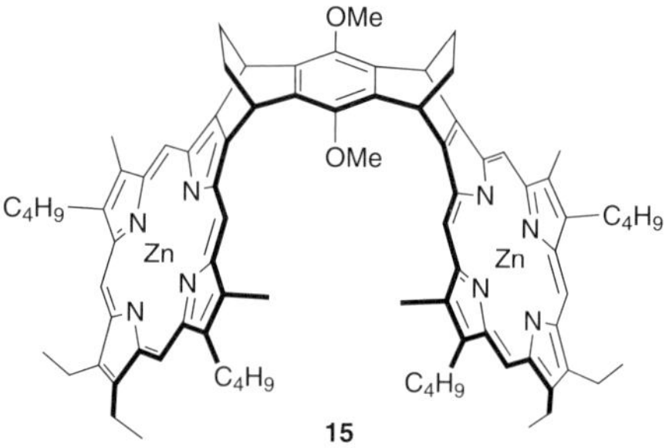

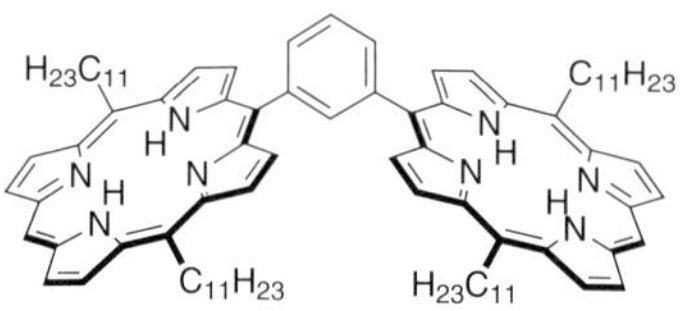

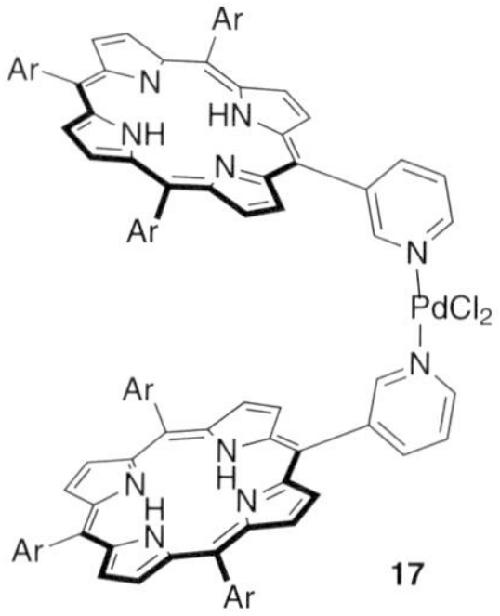

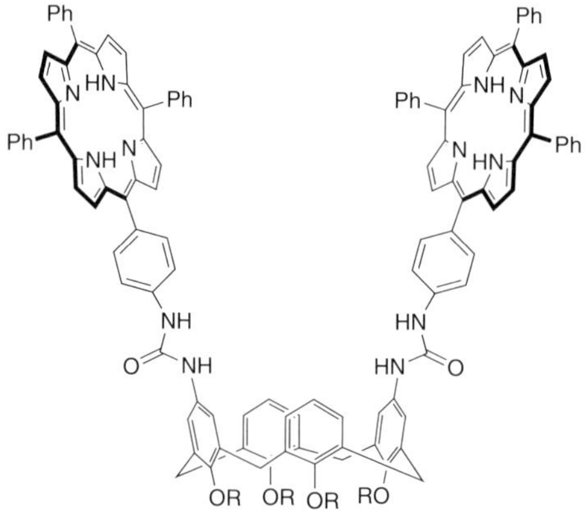

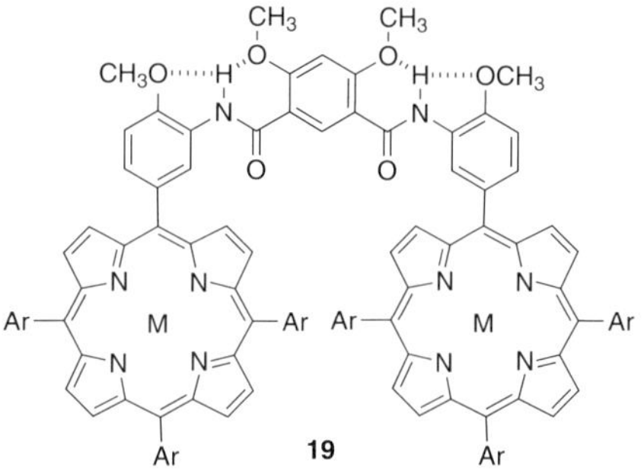

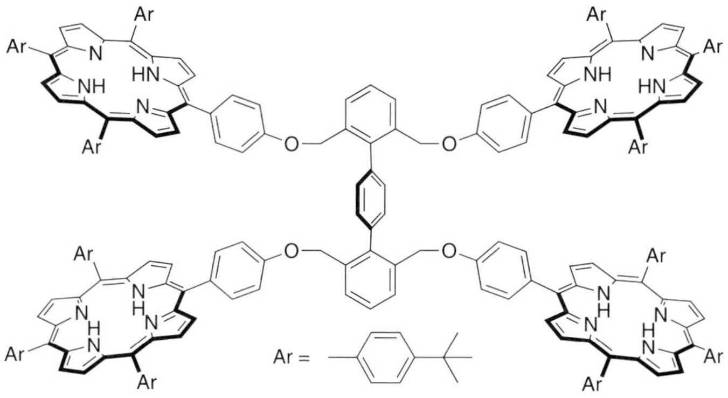

20

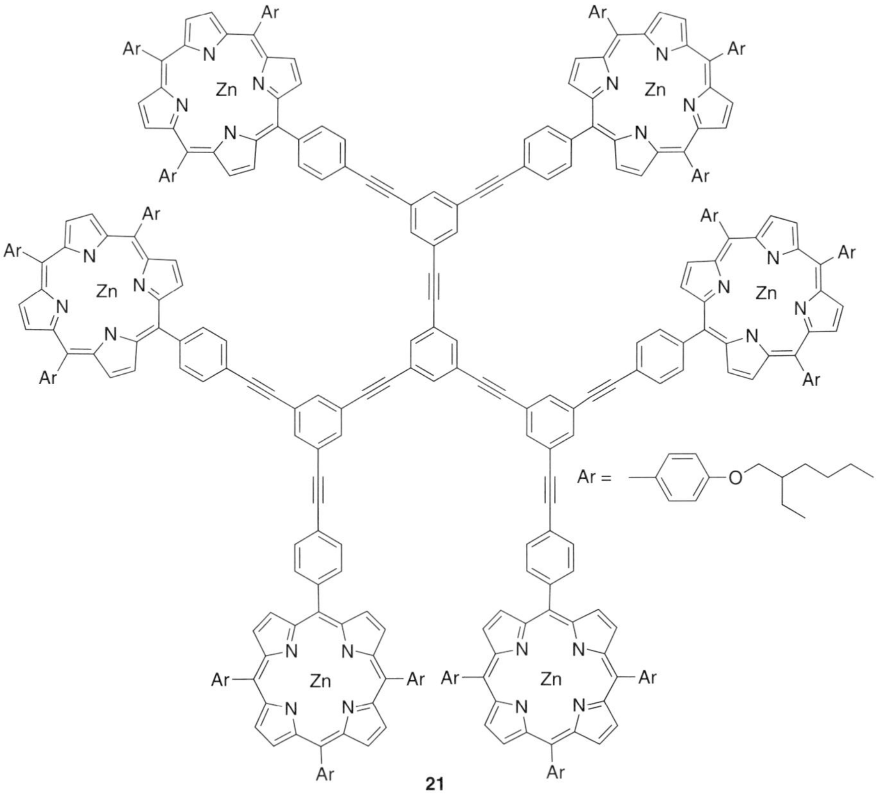

21

Fig. 8 Structures of porphyrin tetramer **20** [104] and hexamer **21** [105]

2.2
Supramolecules of Porphyrins and the Analogues with Carbon Nanotubes

2.2.1
Porphyrin Monomers and the Analogues

Very good affinity between the curved and flat π-surfaces was found in the complexes of fullerenes with porphyrins and the related compounds as described in Sect. 2.1. It is quite natural that this non-classical interaction was applied to CNTs to form complexes with porphyrins and related compounds. The complexation has mainly the following two goals; one is the solubilization of CNTs to make their manipulation much easier, and the other is the investigation of the photo- and electrochemical properties of the complexes [107].

On the other hand, the non-covalent chemistry of CNTs was started in 1998, when cut CNTs were dissolved with the help of surfactants [108] and by wrapping them with highly conjugated macromolecules [109]. Carboxylic groups at the open ends of shortened SWNTs were functionalized covalently with octadecylamine through an amide linkage, also giving SWNT solutions in 1998 [110]. These covalent and non-covalent functionalizations facilitated the processibility of CNTs, making significant technological progress.

As for the non-covalent functionalization with azamacrocyclic molecules, tetraazaannulenes **8** [111] (Fig. 6) and phthalocyanine **22** [112] and porphyrins **23–25** [113] (Fig. 9) were found to form complexes with CNTs. Since the complexes of **8** and **22** did not have enough solubility, they were extracted in the solid phase of CNTs. However, protoporphyrins IX **23–25** dissolved HiPco SWNTs, which were prepared by Carbon Nanotechnologies Co. in the presence of an iron catalyst under high CO pressure and high temperature, in DMF by Nakashima et al. The "DMF dispersion/solution" of SWNTs was obtained after centrifugation at 1000 g for an hour and was very stable for 2 months at 5 °C. Soon after the report, dissolution of SWNTs in a selective manner was reported by Sun with free-base porphyrin monomer **26** [114] (Fig. 9). Non-covalent interaction of SWNTs, prepared by arc discharge, with **26** resulted in the enrichment of semiconducting SWNTs in the solubilized sample and metallic SWNTs in the residual sample, while no SWNTs were solubilized by the *Zn*-analogue **27** (Fig. 9). In an investigation by the author's group, on the other hand, porphyrin monomers highly functionalized with long alkyl chains, **28**, **29** and **30** (Fig. 9), were not able to retain SWNTs, prepared by arc discharge, in THF after centrifugation at 50 400 g for 15 min (Komatsu et al., 2007, unpublished results). Free-base porphyrin **28** was found to show no solubilizing ability of SWNTs under similar conditions as those reported by Li et al. [114]. Therefore, we conclude that it is not so easy to reproduce the result of discriminating semiconducting and metallic SWNTs through the extraction with porphyrin (Komatsu et

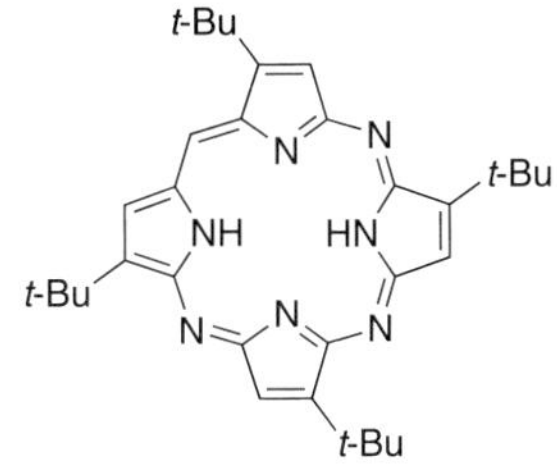

tetrakis(*tert*-butyl)phthalocyanine **22**

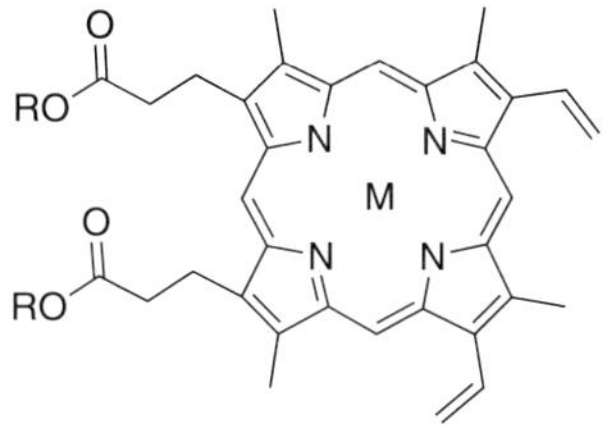

protoporphyrin IX

23; R = H, M = Zn
24; R = CH_3CH_2-, M = H_2
25; R = H, M = FeCl

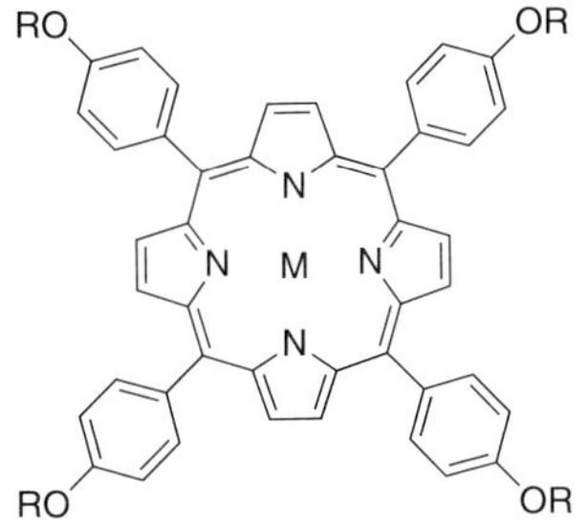

26; R = -$(CH_2)_{15}CH_3$, M = H_2
27; R = -$(CH_2)_{15}CH_3$, M = Zn
28; R = -$(CH_2)_{17}CH_3$, M = H_2
29; R = -$(CH_2)_{17}CH_3$, M = Zn

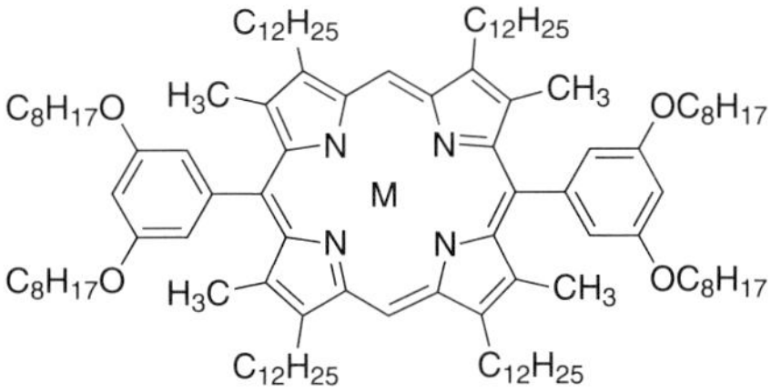

30; M = H_2 and Zn

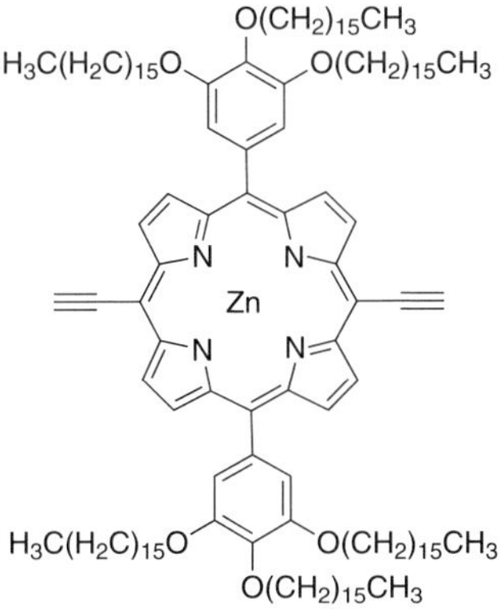

31

Fig. 9 Structures of porphyrin monomers **22–31** [112–115] (Komatsu et al., 2007, unpublished results)

al., 2007, unpublished results). It was also reported that porphyrin monomer **31** (Fig. 9) provided a homogeneous solution of HiPco SWNTs by means of sonication, but the SWNTs precipitated after a few minutes [115]. Very low solubilities (0.1 and 0.08 mg ml^{-1}) of HiPco SWNTs in DMF were also obtained through complexation with **1** (M = Zn and H_2 in Fig. 5), respectively [116]. Compared to the porphyrin oligomers and polymers mentioned below, the interactions between the porphyrin monomer and CNTs is concluded to be much weaker [115], and hence solubilization may be highly dependent on the conditions. It was demonstrated that the stability of the complex between SWNT and porphyrin is proportional to the number of porphyrin units in the triply fused *Zn*-porphyrin oligomers, which will be discussed in Sect. 2.2.2 [117].

Ionic porphyrin monomers and related compounds non-covalently functionalized CNTs and the resulting complexes were dissolved in water and organic solvent. Tetrakis(4-sulfonatophenyl)porphyrin **32** (Fig. 10) was employed to solubilize HiPco SWNTs in water, providing a very stable aqueous solution for several weeks [118]. Protonation of the free-base porphyrin **32** to the diacid under mildly acidic conditions was found to inhibit the interaction between the porphyrin and SWNT. An ionic expanded porphyrin named Sapphyrin **33** (Fig. 10) bound non-covalently to HiPco SWNTs and dissolved them into not only water but also ionic liquid [119]. The supramolecular assemblies of **33** and SWNTs were found to undergo photoexcited intramolecular electron transfer, indicating that the Sapphyrin-bound SWNTs acted as antennae for light-harvesting. In contrast to the case of **32**, the diacidic form of bis(3.5-di-*tert*-butylphenyl)porphyrin **34** (Fig. 10) showed improved ability to solubilize SWNTs (Nanocs Inc.) and construct an ordered supramolecular assembly than that of the unprotonated form of **34** [120]. J- and H-type aggregation effects of the protonated porphyrins on SWNTs played an important role in constructing not only ordered supramolecular assembly at the molecular level, but also large rod-like structures (40–60 nm in diameter and 0.5–3.0 μm in length) at the microscopic level. The photoexcited electron transfer between the SWNTs (Nanocs Inc.) and the protonated forms of the porphyrins, **1** (M = H_2 in Fig. 5), **34** and **35** (Fig. 10), was also observed as in the case of **33** mentioned above [121]. These photochemical behaviors in the supramolecular assemblies were applied to solar cell systems, giving ∼13% of the incident photon to photocurrent efficiency (IPCE) [121].

A similar photochemical behavior was also reported in the nanohybrids of SWNTs (HiPco)/**1** (M = H_2 and Zn in Fig. 5) [116]. The porphyrin-based complexes of SWNTs **36** and C_{60} **37** (Fig. 11) were compared in their photochemical behaviors [122]. In the former complex **36**, SWNTs were doubly functionalized with alkyl chains and 4-pyridylisooxazoline rings at the tips and sidewall to provide sufficient solubility in organic solvent and to coordinate the zinc porphyrin to the pyridyl group, respectively. The covalently functionalized SWNTs formed a complex with a zinc tetraphenylporphyrin in a similar manner to that of C_{60} analogue **37**. In contrast to the photochemical behavior in **37**, energy transfer quenching, rather than electron transfer giving charge-separated states, occurred upon the photochemical excitation of the SWNT–porphyrin hybrid **36**.

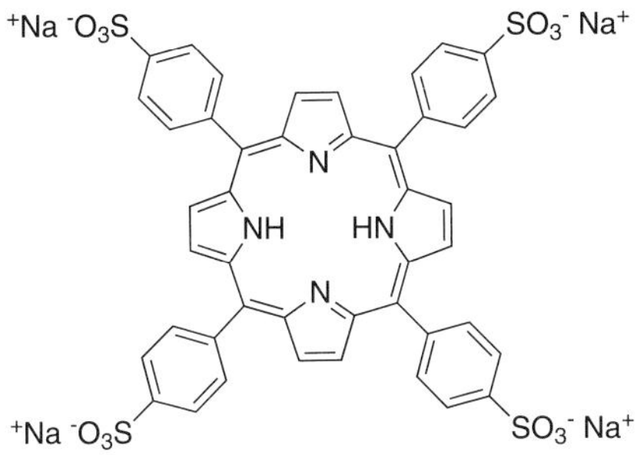

tetrakis(4-sulfonatophenyl)porphyrin **32**

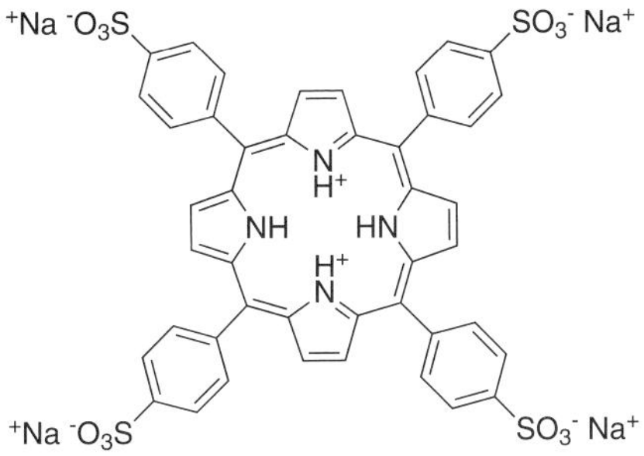

diacidic form of **32**

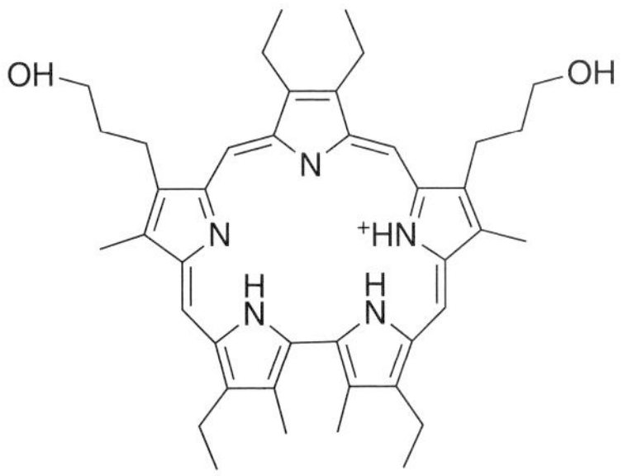

sapphyrin **33**

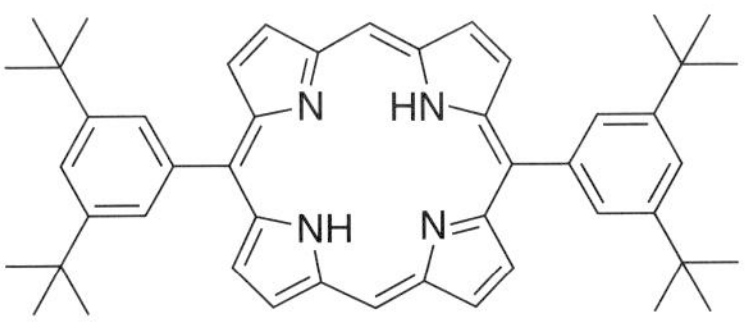

bis(3,5-di-*tert*-butylphenyl)porphyrin **34**

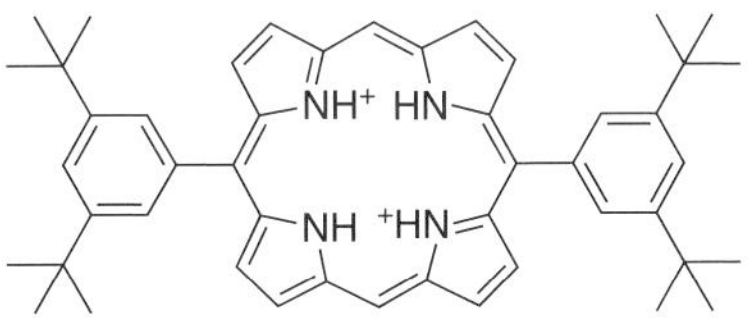

diacidic form of **34**

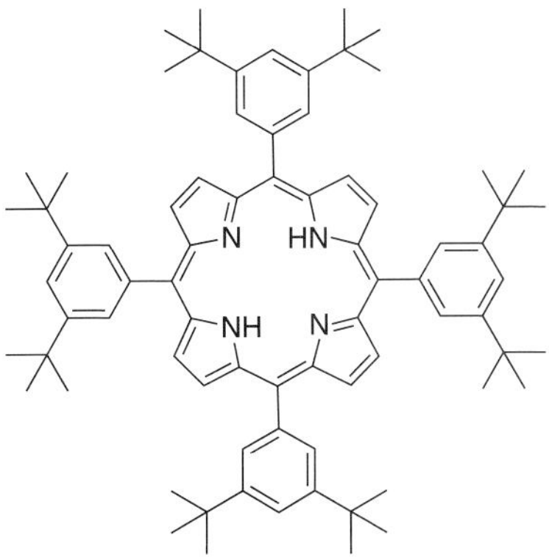

tetrakis(3,5-di-*tert*-butylphenyl)porphyrin **35**

Fig. 10 Structures of ionic porphyrin monomers **32–35** [118–121]

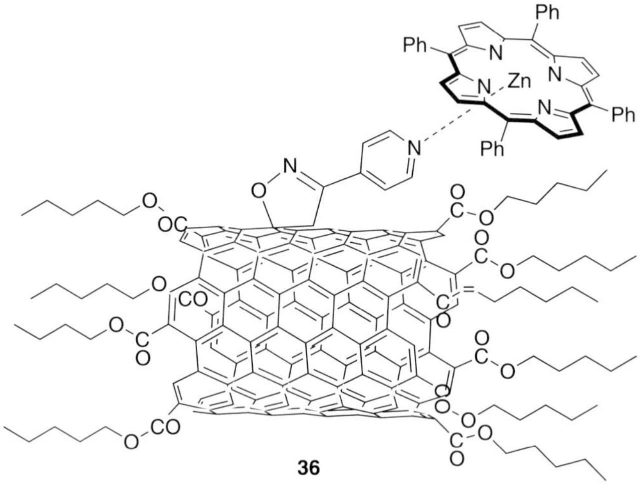

Fig. 11 Structures of porphyrin-based complexes of SWNTs **36** and C_{60} **37** [122]

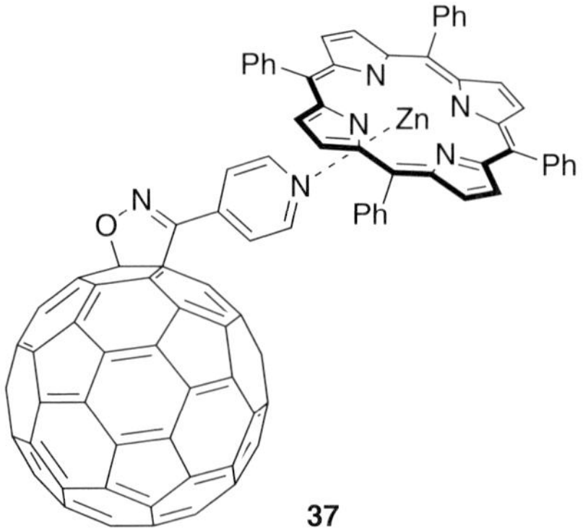

Nanohybrids consisting of three non-covalently linked components were also prepared as shown in Scheme 1 to investigate their photophysical and electrochemical properties [123–125]. HiPco SWNTs and the pyrene ammonium cation were non-covalently bonded through π–π interaction according to the literature [126], giving water-soluble dyad **38**. The anionic porphyrin **39** was immobilized on the dyad **38** through electrostatic interaction, providing the triad **40**. Cationic porphyrin **41** was not able to form a non-covalently linked triad because of the repulsive electrostatic interaction between ammonium and pyridinium. The photophysical and electrochemical studies revealed that electron-transfer from the photoexcited porphyrin **39** to SWNT occurred to form long-lived radical ion pairs (i.e., microseconds).

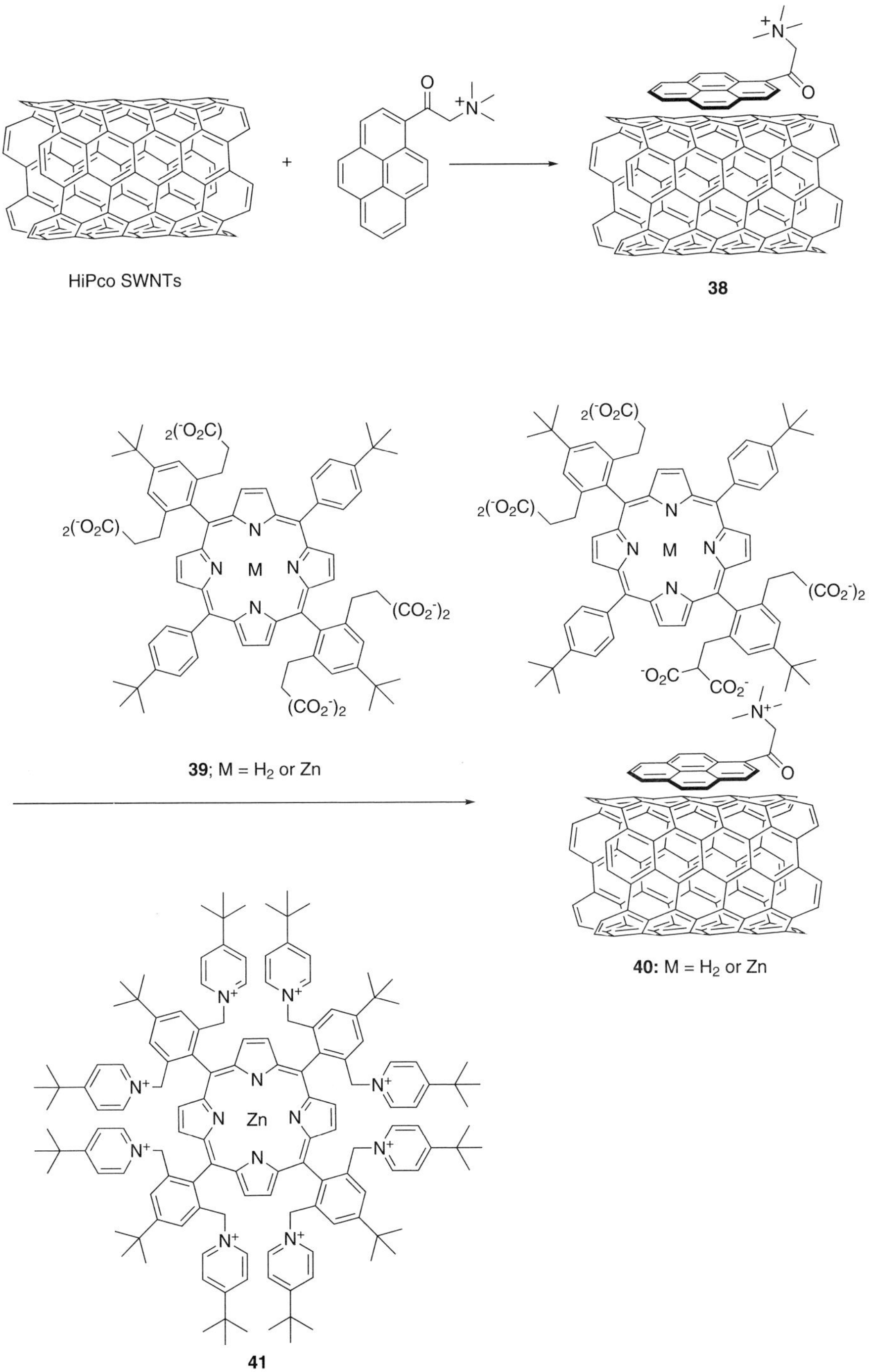

Scheme 1 Preparation of non-covalently linked triad **40** [123–126]

2.2.2
Porphyrin Oligomers and Polymers

As described in Sect. 2.2.1, a more stable complex of CNT is expected to form with the oligomeric and polymeric porphyrins such as **43–45** (Fig. 12) com-

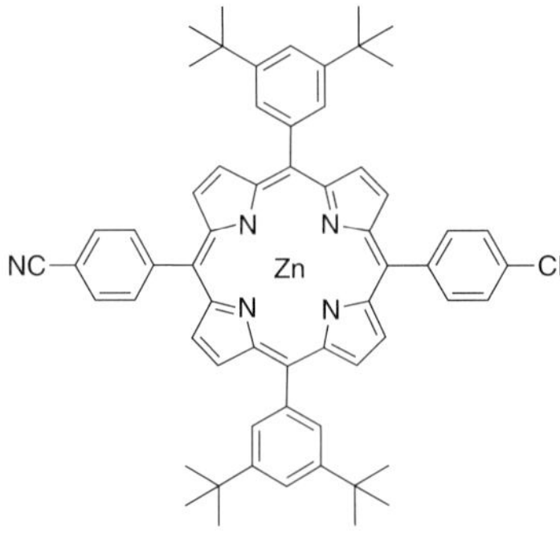

42

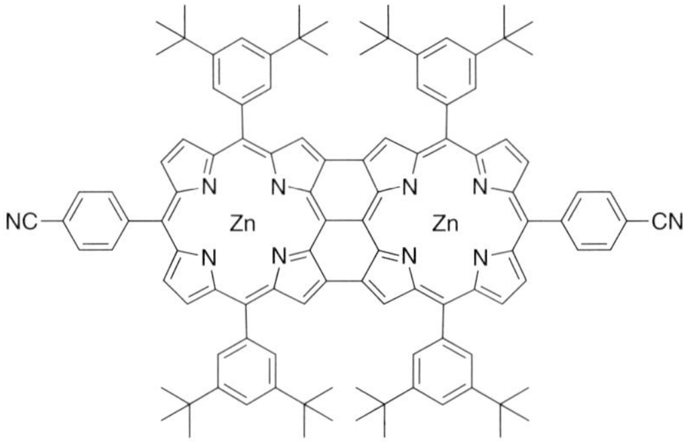

43

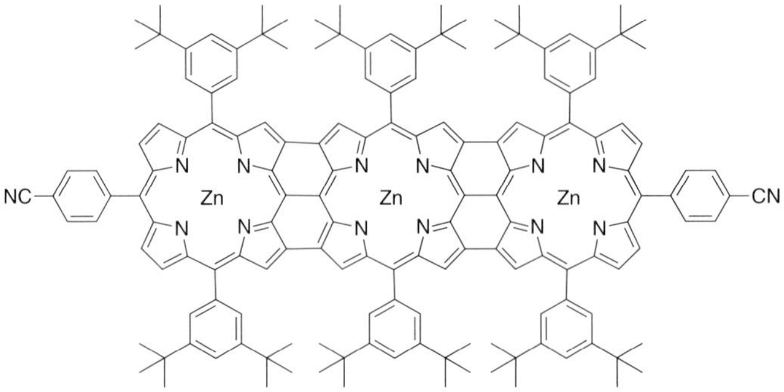

44

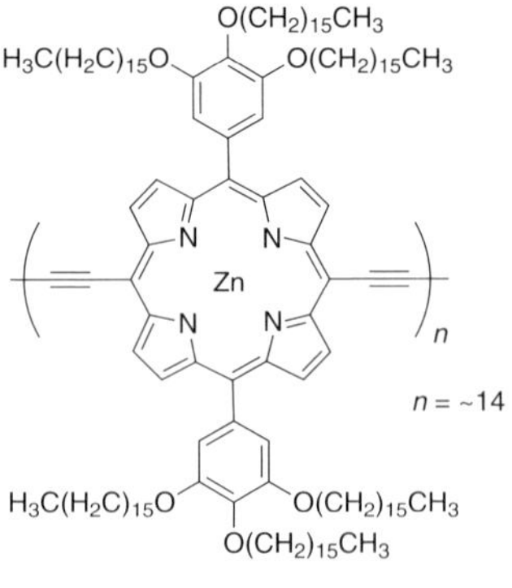

45

Fig. 12 Structures of porphyrin monomer **42**, dimer **43**, trimer **44** and polymer **45** [115, 117] pared to the monomeric ones like **31** (Fig. 9) and **42** (Fig. 12). Actually, the order of the stability was found to be trimer **44** > dimer **43** > monomer **42**, when the triply fused porphyrin trimer and dimer, and monomer were compared in the extraction of HiPco SWNTs in the acidified THF [117]. Trimer **44** gave a very stable, dark solution with no sedimentation even after centrifugation. When the same procedure was employed, no SWNTs were extracted with the monomer **42**, and the SWNTs extracted with dimer **43** precipitated after several hours. Porphyrin monomer **31** and its polymer **45** ($n = \sim 14$) also showed large differences in stability of the SWNT complexes in their solutions; a very dark and stable solution including HiPco SWNTs (approximately 1 mg ml^{-1}) was obtained using **45**, while SWNTs precipitated out soon after sonication in the presence of **31** [115].

Gable-type porphyrin dimers **46** were found to have the ability not only to solubilize SWNTs, but also discriminate their structures [56–58]. These porphyrins consisting of a rigid spacer and the two porphyrin units were prepared via the Suzuki–Miyaura coupling reaction as shown in Scheme 2. In addition, we introduced the asymmetrical moieties at the periphery of the porphyrin units. The chiral porphyrins **46** were found to extract CoMo-

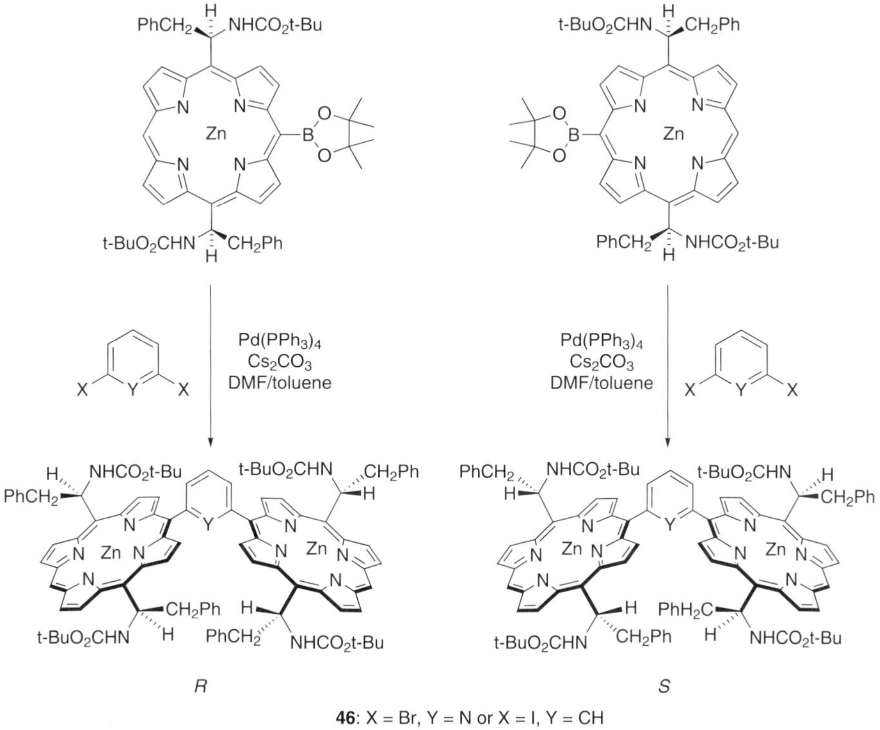

46: X = Br, Y = N or X = I, Y = CH

Scheme 2 Synthesis of gable-type porphyrin dimers **46** [56–58]

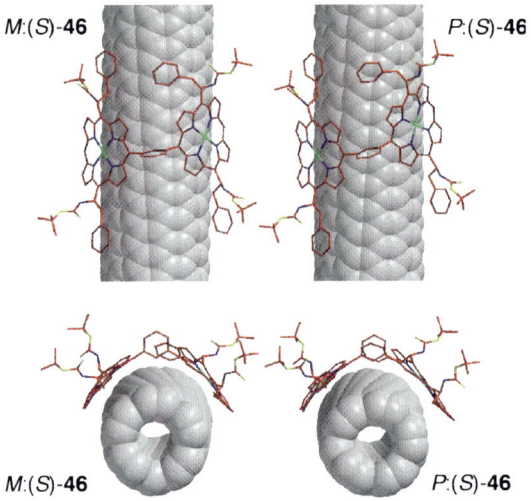

Fig. 13 Computer-generated complex structures of (*M*)- and (*P*)-(6,5) SWNT with (*S*)-**46** [56–58]

CAT SWNTs, prepared by SouthWest NanoTechnologies, Inc. using a CVD method with Co–Mo catalysts, and discriminate the helical arrangements of the hexagons in chiral SWNTs. Computer-generated complex structures between the stereoisomers of (6,5) SWNTs and (*S*)-**46** (Y = CH) are illustrated in Fig. 13. This resulted in obtaining optically active SWNTs for the first time.

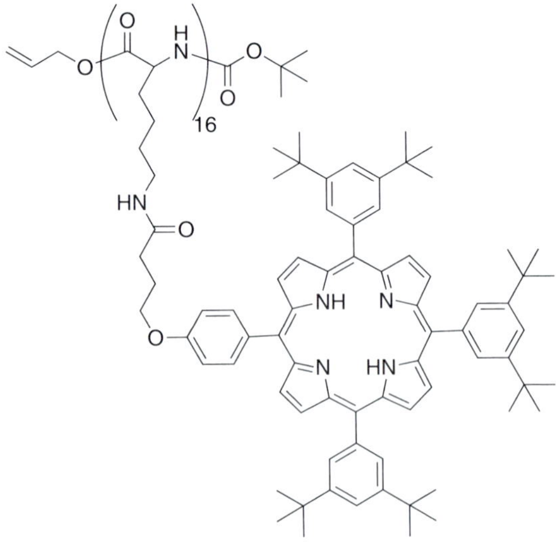

Fig. 14 Structure of porphyrin-peptide hexadecamer **47** [127]

By a similar strategy utilizing strong non-covalent interaction between porphyrin and CNT, diameter-selective extraction was accomplished using porphyrin-peptide hexadecamer **47** [127] (Fig. 14). As in the case with porphyrin dimer **46** [56–58], **47** was removed by washing the complex with DMF and THF several times, enabling us to compare the spectra before and after extraction without any influence of the solubilizing agent. The porphyrin-peptide octamer [128, 129] did not form a complex with HiPco SWNTs, indicating that the long polypeptidic chain played an important role in forming the stable supramolecule by wrapping the circumference of HiPco SWNTs [130]. From Raman spectra of SWNTs before and after extraction, it was found that larger diameters (>1.2 nm) were enriched and the ratio between semiconducting and metallic SWNTs was not influenced so much by

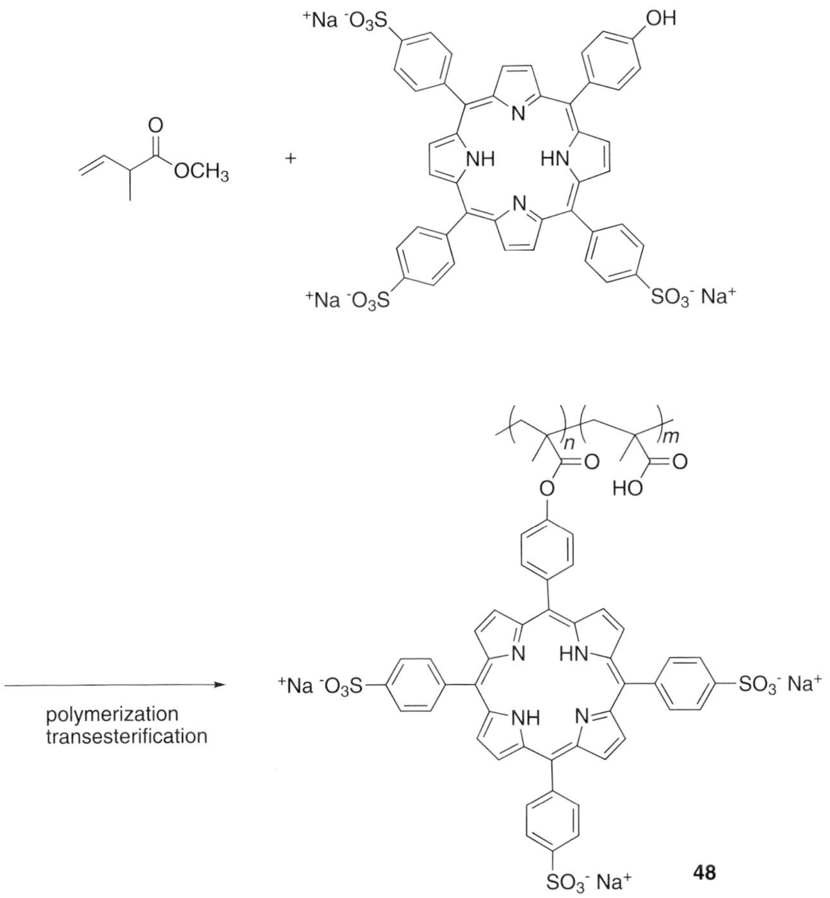

Scheme 3 Preparation of porphyrin/metacrylic acid polymer **48** [131]

the extraction with **47**. In the photochemistry of the supramolecule, a long-lived charge-separated state was attained upon photoexcitation.

Porphyrin-methacrylic acid polymer **48**, prepared by polymerization of methyl methacrylate and transesterification as shown in Scheme 3, was also reported to form a stable complex with SWNTs in DMF through polymer wrapping [131]. The complex was found to create a long-lived charge-separated state upon photoexcitation.

Although the supramolecules of SWNTs wrapped with porphyrin-containing polymers mentioned above were prepared by mixing SWNTs and the polymer in solution, a very stable complex of porphyrin-pyrene copolymer **49**/SWNT was also synthesized by copolymerization of porphyrin and pyrene in the presence of soluble SWNTs as shown in Scheme 4 [132].

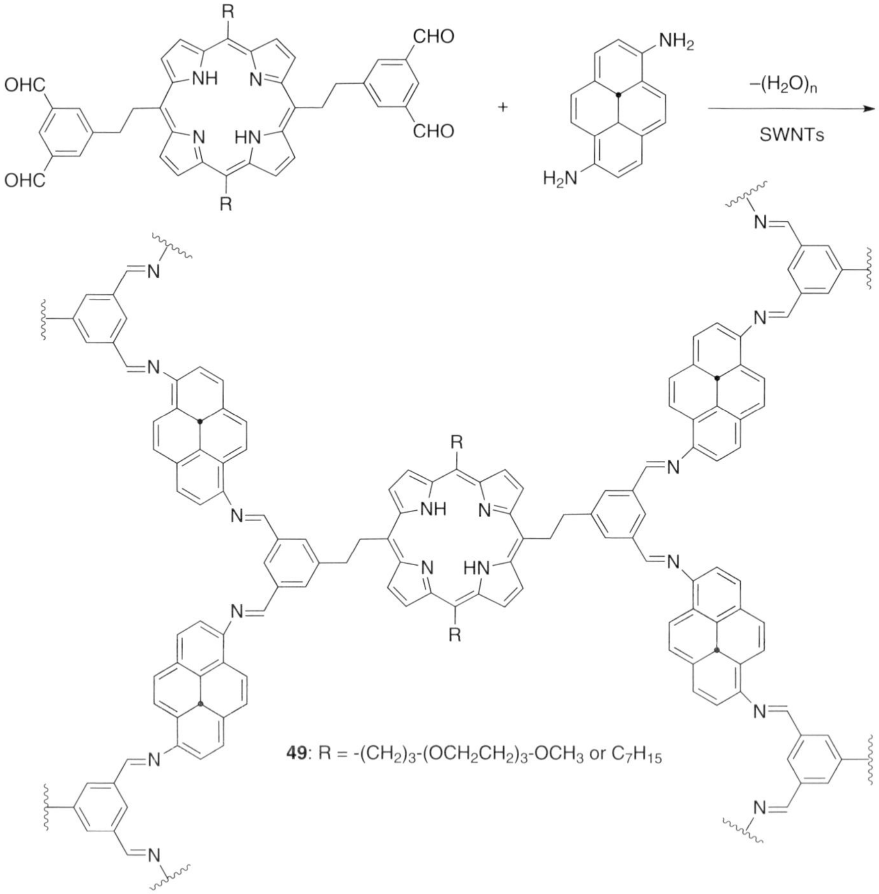

Scheme 4 Preparation of porphyrin/pyrene polymer **49** in the presence of soluble SWNTs [132]

2.3
Supramolecules of DNA with Carbon Nanotubes

Non-covalent immobilization of platinated and iodinated DNA fragments onto the CNT surface was first visualized with HRTEM by Sadler et al. in 1997 [133, 134]. In 2003, DNA was found to be a very efficient solubilizing agent for SWNTs [135–137]. SWNT/DNA hybrids were individually dispersed by helical wrapping of DNA on the circumference of SWNT, giving very stable aqueous solutions under any circumstances. Since then, DNA has been used as a very convenient tool for solubilizing CNTs. The resulting aqueous solution was used for a variety of purposes from biochemical to photo- and electrochemical applications [138].

One of the most successful examples is chromatographic separation of SWNTs wrapped with DNA [136, 137, 139, 140]. Quite recently, (6,4), (9,1) and (6,5) SWNTs were obtained in almost pure form by sorting of CoMoCAT SWNTs wrapped with DNA with conventional ion-exchange chromatography [33]. The SWNTs/DNA solution was subjected to a variety of optical analyses such as circular dichroism, photoluminescence and absorption spectroscopies [141–143] as well as investigation of their photo- and electrochemical behaviors [144, 145]. The strong interaction between SWNT and DNA was simulated theoretically by molecular dynamics [146] and ab initio calculations [147], and used for nanofabrication of SWNTs [148, 149].

Many biological applications have been also examined by CNT/DNA hybrids because of their high chemical stability and biocompatibility [150–152]. In 2004, Bianco et al. reported for the first time that covalently functionalized water-soluble SWNTs were able to cross the cell membrane, showing the great promise of CNTs as a cell-penetrating transporter [153]. Soon after the publication, ammonium-functionalized SWNTs **50** and MWNTs **51** (Fig. 15) were shown to work as a DNA carrier into mammalian cells [154]. In addition, the CNTs exhibited low cytotoxicity, the cytotoxicity being lower than that of other nanomaterials. In order to establish CNT-based gene-transfer vector systems, the interaction of the three types of ammonium-functionalized CNTs **50–52** (Fig. 15) with plasmid DNA and CpG oligodeoxynucleotides was investigated [155, 156].

Kam et al. devised strategies to release biological cargos from non-covalently linked CNT-biomolecule conjugates [157, 158]. The SWNTs were non-covalently functionalized with designed molecules consisting of a hydrophobic phospholipid (PL) moiety to interact with the SWNT surface, a hydrophilic PEG moiety to add aqueous solubility to the hybrids and an X moiety to impart biological function, namely PL-PEG-X **53** (Fig. 16). This "smart nanomaterial" successfully transported and released DNA and siRNA (small interfering RNA) in mammalian cells [158]. This strategy was also applied to selective cancer cell destruction by heating the HiPco SWNT-folic acid conjugate selectively uptaken by cancer cells with near-infrared light [157].

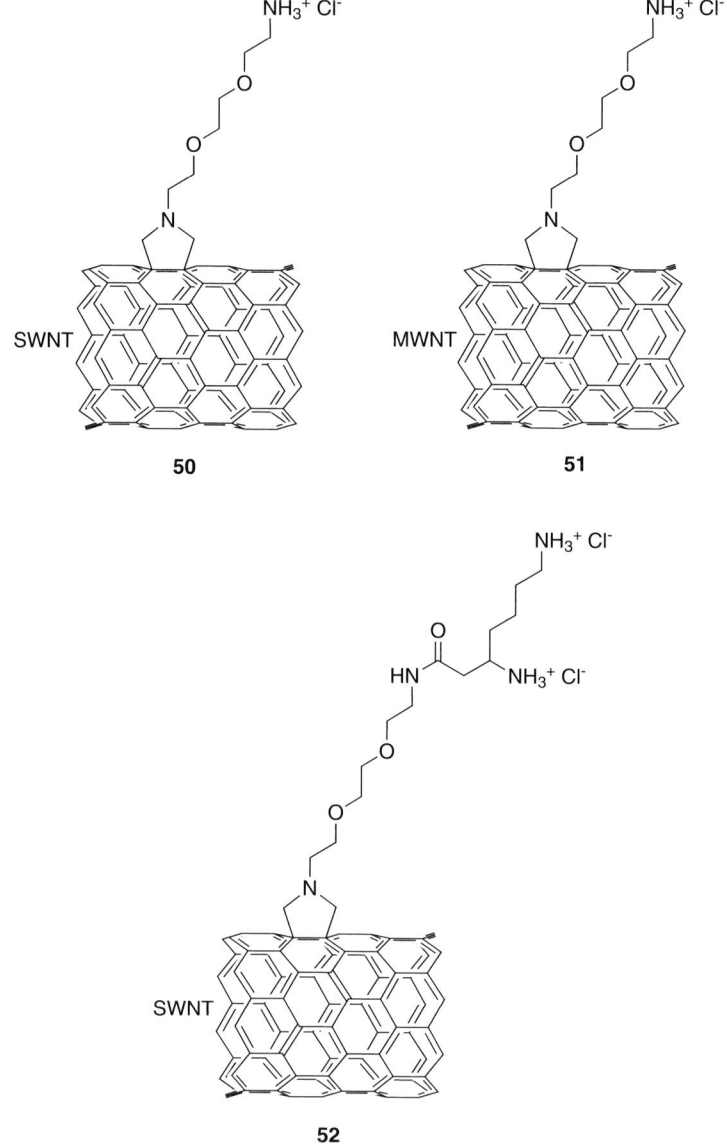

Fig. 15 Structure of ammonium-functionalized CNTs **50–52** [154–156]

CNTs and functionalized CNTs were also applied to DNA sensors [138, 159–161].

DNA (2823–48 502 base pairs) was observed to transport through a MWNT channel directly with fluorescence microscopy [162]. The dynamic process of encapsulating single-stranded DNA (eight adenine bases) in (10,10) SWNT was simulated in a water solute environment, indicating spontaneous inser-

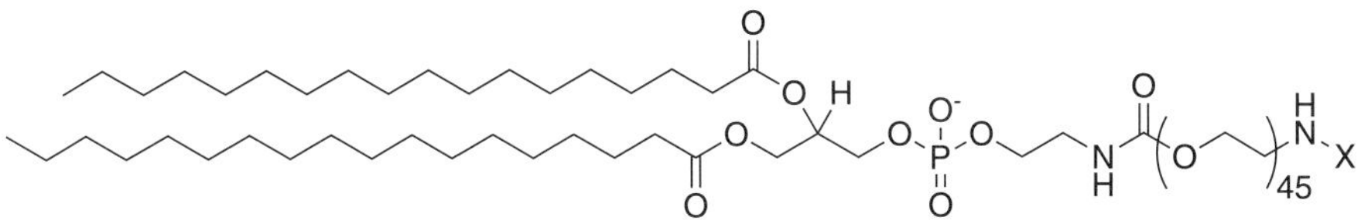

Fig. 16 Structure of PL (phospholipids)-PEG-X **53** [157, 158]

tion and confinement of the DNA inside the SWNT by van der Waals and hydrophobic forces [163].

2.4
Supramolecules of Proteins with Carbon Nanotubes

As with DNA above, a variety of proteins were immobilized non-covalently not only outside but also inside the CNT [138, 150]. Tsang et al. reported the first evidence for immobilization of small proteins inside MWNT by HRTEM [164–167], before DNA immobilization on the surface of MWNT as mentioned above [133, 134]. After the innovative work by Sadler et al., the circumference of CNTs was non-covalently functionalized with proteins directly or indirectly. MWNTs, prepared by arc discharge, were shown to be almost completely covered by streptavidin in helical arrangement [168]. SWNTs also adsorbed proteins and enzymes such as cytochrome c, ferritin, glucose oxidase and anti-fullerene IgG monoclonal antibody [169, 170]. Indirect non-covalent functionalizations of SWNTs with streptavidine and metallothionein were attained through triton-X 100-PEG-biotin and butylpyrene linkers, respectively [171, 172].

CNTs transported proteins into cells in a similar manner to that of DNA- and peptide-functionalized CNTs mentioned above [153, 154]. SWNTs non-covalently conjugated with various proteins such as streptavidine, cyctochrome c, protein A and bovin serum albumin were investigated for carriage into mammalian cells like HeLa, NIH-3T3 fibroblast, HL60 and Jurkats cells [173, 174].

Non-covalently functionalized SWNTs were also applied to highly specific electronic and optical protein biosensors [138, 175–178].

2.5
Supramolecules of Peptides with Carbon Nanotubes

The covalent chemistry of peptide-functionalized CNTs has made significant progress mainly in the biomedical applications mentioned above [151–153]. In contrast, non-covalent peptide chemistry of CNTs has been rather limited. In 2003, peptide sequences with specific affinity for various kinds of CNTs,

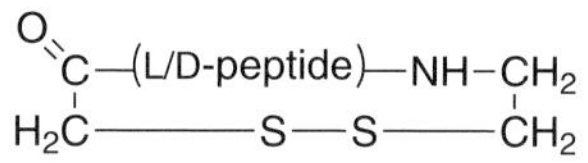

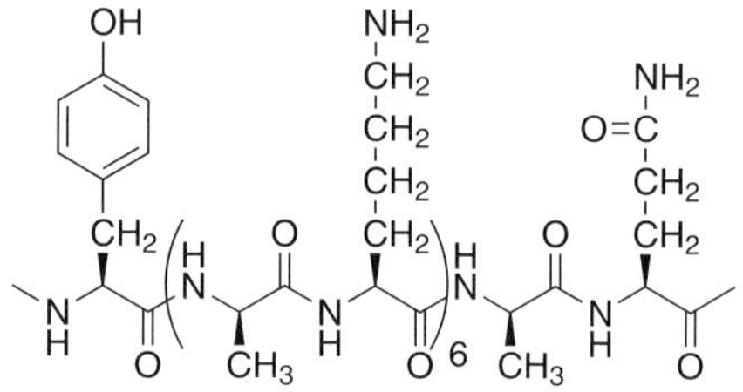

Fig. 17 Structures of cyclic peptides **54** and **55** [130, 182]

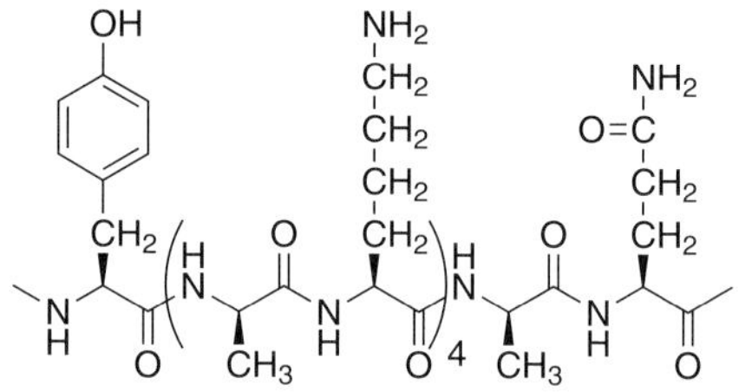

prepared by laser vaporization, arc discharge, and HiPco and CVD processes, were investigated, showing that sequences rich in histidine and tryptophan acted in unison for binding CNTs [179]. A 29-residue peptide designed to form an amphiphilic α-helix was found to solubilize HiPco SWNTs and control the assembly of peptide-coated SWNTs through the adjacent peptide–peptide interactions [180, 181]. When the cyclic peptides, **54** and **55** (Fig. 17), were employed for the solubilization of HiPco SWNTs, smaller diameters were enriched in the aqueous supernatant [130, 182].

2.6
Supramolecules of Carbohydrates with Fullerenes

It is well known that fullerenes show good solubility in many organic solvents such as toluene, carbon disulfide and dichlorobenzene, allowing significant progress in the chemistry of fullerenes. On the other hand, the solubilization of fullerenes in water is also of great importance in particular for their biomedical applications. Although there are a lot of examples to achieve water-soluble fullerenes by covalent functionalization, non-covalent functionalization with CDs is an efficient and convenient way to solubilize fullerenes in water without any loss of the π-conjugation [15–17, 183].

In 1992, a water-soluble fullerene was first prepared by Anderson et al. via non-covalent functionalization of C_{60} with γ-CD **58** [184] (Fig. 18). The com-

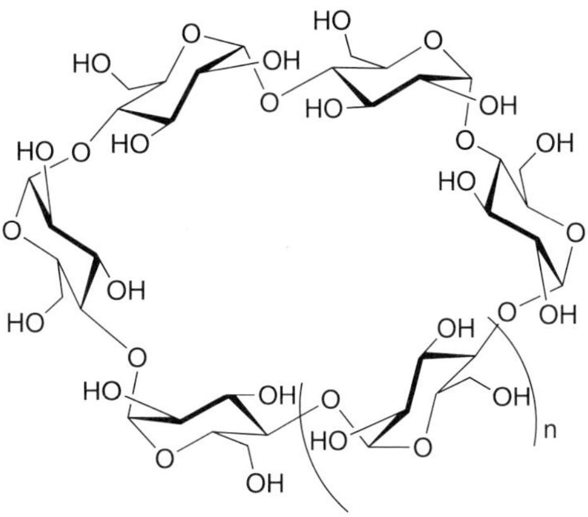

Fig. 18 Structures of cyclodextrins **56–59** [141, 184–203]

56: α-CD (n = 1)
57: β-CD (n = 2)
58: γ-CD (n = 3)
59: η-CD (n = 7)

plex was analyzed with NMR, UV-VIS and FAB-MS spectroscopies [185, 186]. A similar procedure was applied to the synthesis of a water-soluble C_{70}-γ-CD complex [187]. Changing the solvent from water to water/toluene and DMF/toluene mixed solvent systems provided C_{60}-γ-CD [188] and C_{60}-β-CD **57** [189] (Fig. 18) complexes, respectively, with fixed stoichiometry of 1:2, namely bicapped C_{60}. A C_{60}-γ-CD complex was also prepared under milder conditions at room temperature by use of methanol as a solvent [190]. A mechanochemical technique named high-speed vibration milling used for bridging two C_{60} molecules by covalent bonding [191], was also applied to the non-covalent functionalization of C_{60}, C_{70} and C_{60} derivatives with γ-CD [192, 193]. The photo- and electrochemical properties and complex structures of C_{60}-γ-CD were characterized by experiments such as laser photolysis and cyclic voltammetry, and spectroscopic analyses such as UV-Vis, fluorescence and NMR [185, 190, 194–199].

2.7
Supramolecules of Carbohydrates with Carbon Nanotubes

Both cyclic and linear carbohydrates showed good affinity and sufficient solubilizing ability to CNTs as in the case of other biomolecules, DNA, proteins and peptides, mentioned so far [150].

After HiPco SWNTs were ground in the presence of β-CD **57** or γ-CD **58** (Fig. 18), the resulting black powder was found to contain shortened SWNTs and could be dissolved in water [200]. HiPco SWNTs were also dissolved in water by refluxing them in the aqueous γ-CD solution [201] in a similar procedure to that for preparing γ-CD-C_{60} complex [184, 187]. The SWNTs non-covalently functionalized with γ-CD by means of grinding and refluxing were analyzed with absorption and Raman spectroscopies, and differential scanning calorimetry [201]. The mechanochemical high-speed vibration milling technique developed by Braun et al. and Komatsu et al. [191, 192, 202] was also applied to prepare an aqueous solution of SWNT/α-, β- and γ-CD, **56–58** (Fig. 18), complexes as in the case of the fullerene-CD complexes mentioned above [203].

The CD with the larger ring, η-CD **59** (Fig. 18), was employed to thread SWNTs (MER Co., USA), resulting in solubilizing them in water and separating them with respect to diameters [141]. However, more experimental evidence such as Raman, absorption and photoluminescence spectroscopies is considered to be required to confirm firmly the diameter-based separation in the paper by Dodziuk et al. (2003) [141].

Naturally occurring non-cyclic polysaccharides were shown to act as strong solubilizing agents. Starch dissolved HiPco SWNTs in water, and enzymatic hydrolysis of the aqueous solution with amyloglucosidase precipitated out the dissolved SWNTs [204]. Amylose **60** (Fig. 19), one of the major components of starch, wrapped HiPco SWNTs in a helical manner to dissolve the resulting complex into aqueous DMSO [205]. Gum Arabic (GA), a water-soluble polysaccharide produced by *Acaciasenegal* trees, exhibited better solubilizing ability for both SWNTs and MWNTs than those of typical surfactants such as sodium dodecylsulfate (SDS), dextrin and PEG [206]. Diameter-selective dispersion of HiPco SWNTs was reported by use of chi-

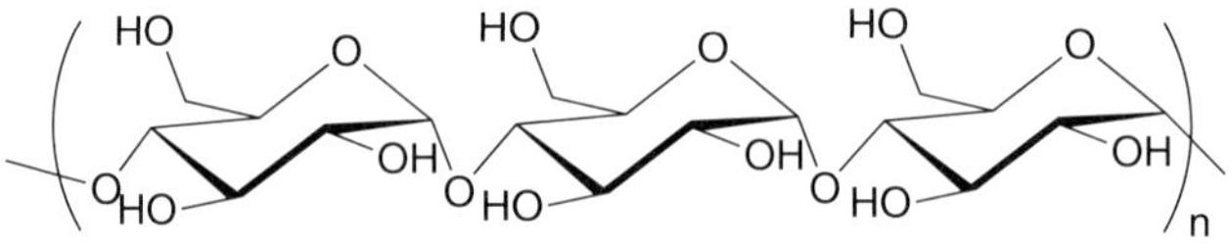

amylose **60**

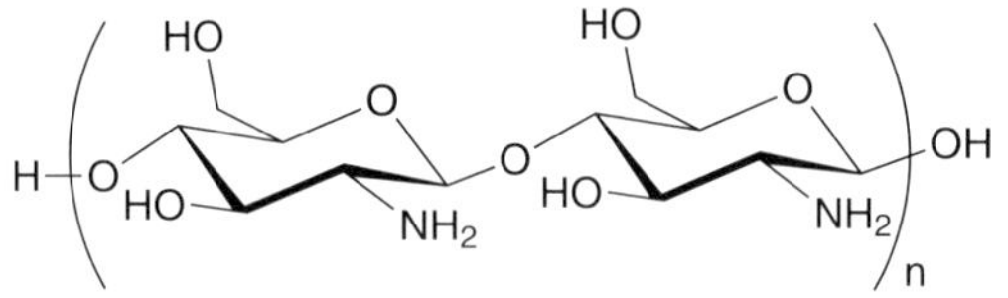

chitosan **61**

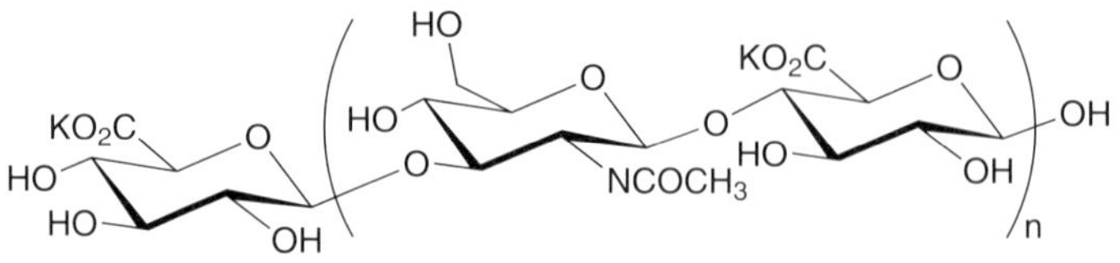

hyaluronic acid **62**

Fig. 19 Structures of polysaccharide **60–62** [205, 207–211]

tosan **61** (Fig. 19) as a solubilizing agent [207]. Since **61** has amino groups, it is expected that **61** will discriminate the structures and electronic properties of SWNTs as demonstrated with simple alkyl amines [208–210]. In fact, smaller-diameters of SWNTs were preferentially dispersed in the aqueous supernatant, while larger-diameters were enriched in the precipitate. Hyaluronic acid **62** (Fig. 19), like other polysaccharides mentioned above, exhibited remarkable dispersive ability to HiPco SWNTs [211]. Because of the organization of SWNTs, a nematic liquid crystal phase was separated in equilibrium with an isotropic phase.

3
Endohedral Supramolecular Chemistry of Fullerenes and Carbon Nanotubes

Encapsulation of metal atoms inside the hollow cage of fullerene was first reported in 1985 [212], immediately after the discovery of C_{60} [1]. Since then, many endohedral metallofullerenes have been prepared and characterized, and many fascinating properties have been disclosed [213–217]. CNTs followed the same history as fullerenes [34, 218]; the hollow core of open CNT was reported to be partially filled with metals, immediately after the discovery and bulk synthesis of CNTs [219–223]. Since the hollow space inside CNTs is larger than that of fullerene, CNTs can encapsulate molecules, which is in contrast to the fact that fullerenes can accept atoms or very small molecules like H_2 at most [10].

Immobilization of small proteins in MWNTs with 3.0–5.0 nm in inner tube diameters was first observed with HRTEM in 1995 as mentioned in Sect. 2.4 [164, 165]. CNTs were non-covalently hybridized with fullerenes to provide "fullerene peapods" in 1998 [49, 224]. This novel nanomaterial attracted considerable attention because of the great possibility for tuning the electronic structures of CNTs and shielding the encapsulated molecules by the carbon cage [222].

Since then, a variety of molecular peas were integrated into SWNTs [34, 225]; the molecules with three-dimensional structures are endohedral metallofullerenes [226–229], metallocenes [230, 231] and o-carboranes [232, 233]. Molecular and/or atomic motions in the confined space inside the tube were successfully observed by HRTEM due partly to the shielding effect of the carbon cage from electron impact [227–229, 232, 233]. Two- and one-dimensional conjugated molecules were also encapsulated in SWNTs [234–237]. Relatively small molecules such as tetramethyltetraselenafulvalene **63**, tetrathiafulvalene **64**, tetracyanoquinodimethane **65** and tetrafuluorotetracyanoquinodimethane **66** (Fig. 20) were encapsulated in SWNTs to modify their electronic structures [238]. Ionic liquid 1-butyl-3-methylimidazolium hexafluorophosphate [bmim][PF_6] **67** (Fig. 20) was found to change the physical properties when it was confined in MWNTs [239].

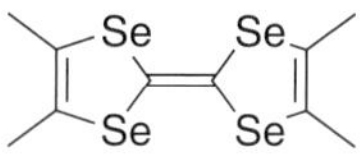

Fig. 20 Structures of heterocyclic peas in peapods **63–68** [238–240]

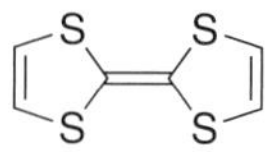

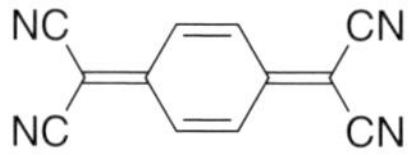

Zn-diphenylporphyrin **68** (Fig. 20) and analogues, which are much larger molecules than **63–67**, were found to be encaged inside SWNTs [240, 241]. As described in Sect. 2.3, DNA was inserted in and even transported through CNT [162, 163]. Linear polyyne molecules, $C_{2n}H_2$, were highly stabilized by encapsulating in SWNTs even above 300 °C under dry-air conditions [237, 242].

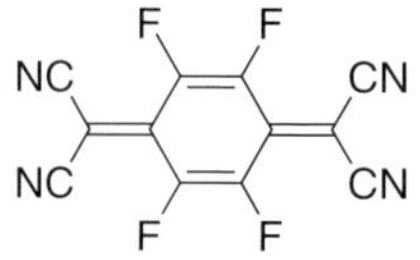

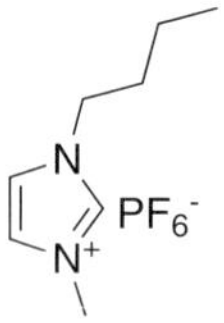

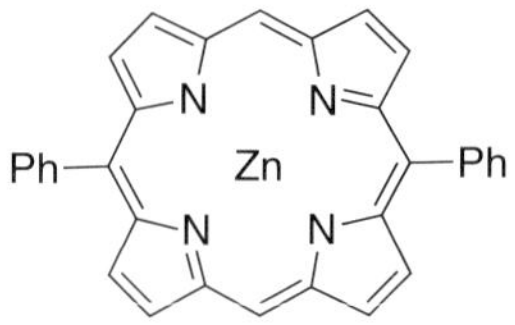

4
Concluding Remarks

This review has described the supramolecular chemistry of fullerenes and CNTs. Although the non-covalent chemistry of heterocyclic molecules has been highlighted in this review, the supramolecular chemistry of other host molecules, as well as covalent chemistry, has also contributed enormously to the progress of this interdisciplinary field in science. One of the landmarks must be the application of the supramolecular interaction of C_{60} with calix[8]arenes to the practical purification of fullerenes, which was reported independently by Shinkai et al. and Atwood et al. in 1994 [243, 244]. Quite recently, the host–guest strategy has been successfully applied to structure-based separation of CNTs by the author's group [56–58]. These examples and the works reviewed above clearly ensure that supramolecular chemistry will continue to contribute to the progress of nanoscale science and technology. In particular, the hurdles in structural separation, nanofabrication and bioapplications of CNTs will hopefully be addressed by the supramolecular strategy.

Acknowledgements The author thanks Dr. Xiaobin Peng (Shiga University of Medical Science) and Mr. Naoki Kadota (Kyoto University) for help in preparing this review. The author is also grateful to Prof. Atsuhiro Osuka (Kyoto University) and Prof. Takahide Kimura (Shiga University of Medical Science) for their encouragement.

References

1. Kroto HW, Health JR, O'Brien SC, Curl RF, Smalley RE (1985) Nature 318:162
2. Krätschmer W, Lamb LD, Fostiropoulos K, Huffman DR (1990) Nature 347:354
3. Komatsu N, Kadota N, Kimura T, Kikuchi Y, Arikawa M (2007) Fullerenes Nanotubes Carbon Nanostruct 15:217
4. Komatsu N, Ohe T, Matsushige K (2004) Carbon 42:163
5. Nagata K, Dejima E, Kikuchi Y, Hashiguchi M (2005) Chem Lett 34:178
6. Theobald J (1995) Sep Sci Technol 30:2783
7. Murayama H, Tomonoh S, Alford JM, Karpuk ME (2004) Fullerenes Nanotubes Carbon Nanostruct 12:1
8. Scott LT, Boorum MM, McMahon BJ, Hagen S, Mack J, Blank J, Wegner H, Meijere A (2002) Science 295:1500
9. Boorum MM, Vasil'ev YV, Drewello T, Scott LT (2001) Science 294:828
10. Komatsu K, Murata M, Murata Y (2005) Science 307:238
11. Ikeda A, Shinkai S (1997) Chem Rev 97:1713
12. Shinkai S, Ikeda A (1997) Gazz Chim Ital 127:657
13. Shinkai S, Ikeda A (1999) Pure Appl Chem 71:275
14. Lhotak P, Shinkai S (1995) J Synth Org Chem Jpn 53:963
15. Raston CL (1996) In: Atwood JL, Davies JED, MacNicol DD, Vogtle F, Suslick KS (eds) Comprehensive Supramolecular Chemistry, Vol 1. Pergamon, Oxford, p 777
16. Diederich F, Gomez-Lopez M (1999) Chem Soc Rev 28:263
17. Constable EC (1994) Angew Chem Int Ed Engl 33:2269
18. Komatsu N (2001) Tetrahedron Lett 42:1733
19. Komatsu N (2003) Org Biomol Chem 1:204
20. Matsubara H, Hasegawa A, Shiwaku K, Asano K, Uno M, Takahashi S, Yamamoto K (1998) Chem Lett: 923
21. Matsubara H, Oguri S, Asano K, Yamamoto K (1999) Chem Lett: 431
22. Matsubara H, Shimura T, Hasegawa A, Senba M, Asano K, Yamamoto K (1998) Chem Lett: 1099
23. Kawase T, Fujiwara N, Tsutumi M, Oda M, Maeda Y, Wakahara T, Akasaka T (2004) Angew Chem Int Ed 43:5060
24. Kawase T, Seirai Y, Darabi HR, Oda M, Sarakai Y, Tashiro K (2003) Angew Chem Int Ed 42:1621
25. Kawase T, Tanaka K, Fujiwara N, Darabi HR, Oda M (2003) Angew Chem Int Ed 42:1624
26. Kawase T, Tanaka K, Seirai Y, Shiono N, Oda M (2003) Angew Chem Int Ed 42:5597
27. Kawase T, Tanaka K, Shiono N, Seirai Y, Oda M (2004) Angew Chem Int Ed 43:1722

28. Kawase T (2007) J Synth Org Chem Jpn 65:888
29. Sun Y-P, Fu K, Lin Y, Huang W (2002) Acc Chem Res 35:1096
30. Sinnott SB (2002) J Nanosci Nanotechnol 2:113
31. Niyogi S, Hamon MA, Hu H, Zhao B, Bhowmik P, Sen R, Itkis ME, Haddon RC (2002) Acc Chem Res 35:1105
32. Hirsch A (2002) Angew Chem Int Ed 41:1853
33. Zheng M, Semke ED (2007) J Am Chem Soc 19:6084
34. Britz DA, Khlobystov AN (2006) Chem Soc Rev 35:637
35. Iijima S (1991) Nature 354:56
36. Iijima S, Ichihashi T (1993) Nature 363:603
37. Bethune DS, Kiang CH, Vries MSd, Gorman G, Savoy R, Vazquez J, Beyers R (1993) Nature 363:605
38. Monthioux M, Kuznetsov VL (2006) Carbon 44:1621
39. Boehm HP (1997) Carbon 35:581
40. Gibson JAE (1992) Nature 359:369
41. Wildgoose GG, Banks CE, Compton RG (2006) Small 2:182
42. Ando T (2006) New Diamond 22:8
43. Andrews R, Jaques D, Qian D, Rantell T (2002) Acc Chem Res 35:1008
44. Rao CNR, Govindaraj A (2002) Acc Chem Res 35:998
45. Endo M, Muramatsu H, Hayashi T, Kim YA, Terrones M, Dresselhaus MS (2005) Nature 433:476
46. Sugai T, Yoshida H, Shimada T, Okazaki T, Shinohara H (2003) Nano Lett 3:769
47. Flahaut E, Bacsa R, Peigney A, Laurent C (2003) Chem Commun: 1442
48. Bachilo SM, Balzano L, Herrera JE, Pompeo F, Resasco DE, Weisman RB (2003) J Am Chem Soc 125:11186
49. Wang B, Poa CHP, Wei L, Li L-J, Yang Y, Chen Y (2007) J Am Chem Soc 129:9014
50. Ouyang M, Huang J-L, Lieber CM (2002) Acc Chem Res 35:1018
51. Charlier J-C (2002) Acc Chem Res 35:1063
52. Samsonidze GG, Grüneis A, Saito R, Jorio A, Souza Filho AG, Dresselhause G, Dresselhause MS (2004) Phys Rev B 69:205402
53. Sánchez-Castillo A, Román-Velázquez CE, Noguez C (2006) Phys Rev B 73:045401
54. Moss GP (1996) Pure Appl Chem 68:2193
55. Strano MS (2007) Nat Nanotechnol 2:340
56. Peng X, Komatsu N, Bhattacharya S, Shimawaki T, Aonuma S, Kimura T, Osuka A (2007) Nat Nanotechnol 2:361
57. Peng X, Komatsu N, Kimura T, Osuka A (2007) J Am Chem Soc 129:15947
58. Komatsu N (2008) New Diamond 88:26
59. Szabados Á, Biró LP, Surján PR (2006) Phys Rev B 73:195404
60. Liu Z, Suenaga K, Yoshida H, Sugai T, Shinohara H, Iijima S (2005) Phys Rev Lett 95:187406
61. Damnjanovic M, Milosevic I, Vukovic T, Sredanovic R (1999) Phys Rev B 60:2728
62. Kibbey CE, Savina MR, Parseghian BK, Francis AH, Meyerhoff ME (1993) Anal Chem 65:3717
63. Xiao J, Savina MR, Martin GB, Francis AH, Meyerhoff ME (1994) J Am Chem Soc 116:9341
64. Xiao J, Meyerhoff ME (1995) J Chromatogr 715:19
65. Coutant DE, Clarke SA, Francis AH, Meyerhoff ME (1999) In: Jinno K (ed) Separation of Fullerenes by Liquid Chromatography. The Royal Society of Chemistry, Cambridge, p 129

66. Eichhorn DM, Yang S, Jarrell W, Baumann TF, Beall S, White AJP, Williams DJ, Barret AGM, Hoffman BM (1995) J Chem Soc Chem Commun: 1703
67. Sun Y, Drovetskaya T, Bolskar RD, Bau R, Boyd PDW, Reed CA (1997) J Org Chem 62:3642
68. Maruyama H, Fujiwara M, Tanaka K (1998) Chem Lett: 805
69. Olmstead MM, Costa DA, Maitra K, Noll BC, Phillips SL, Calcar PMV, Balch AL (1999) J Am Chem Soc 121:7090
70. Boyd PDW, Hodgson MC, Richard CEF, Oliver AG, Chaker L, Brother PJ, Bolskar RD, Tham FS, Reed CA (1999) J Am Chem Soc 121:10487
71. Amaroli N, Diederich F, Echegoyen L, Habicher T, Flamigni L, Marconi G, Nierengarten J-F (1999) New J Chem: 77
72. Konarev DV, Neretin IS, Slovokhotov YL, Yudanova EI, Drichko NV, Shul'ga YM, Tarasov BP, Gumanov LL, Batsanov AS, Howard JAK, Lyubovskaya RN (2001) Chem Eur J 7:2605
73. Bhattacharya S, Shimawaki T, Peng X, Ashokkumar A, Aonuma S, Kimura T, Komatsu N (2006) Chem Phys Lett 430:435
74. Wang Y-B, Zhenyang L (2003) J Am Chem Soc 125:6072
75. Bhattacharya S, Ujihashi N, Aonuma S, Kimura T, Komatsu N (2007) Spectrochim Acta A 68:495
76. Guldi DM, Ros TD, Braiuca P, Prato M, Alessio E (2002) J Mater Chem 12:2001
77. Sun D, Tham FS, Reed CA, Boyd PDW (2002) PNAS 99:5088
78. Kimura M, Saito Y, Ohta K, Hanabusa K, Shirai H, Kobayashi N (2002) J Am Chem Soc 124:5274
79. Kimura M, Shiba T, Yamazaki M, Hanabusa K, Shirai H, Kobayashi N (2001) J Am Chem Soc 123:5636
80. Ishii T, Aizawa N, Kanehama R, Yamashita M, Sugiura K, Miyasaka H (2002) Coord Chem Rev 226:113
81. Balch AL, Olmstead MM (1999) Coord Chem Rev 185–186:601
82. Boyd PDW, Reed CA (2005) Acc Chem Res 38:235
83. Sun D, Tham FS, Reed CA, Chaker L, Boyd PDW (2002) J Am Chem Soc 124:6604
84. Andrews PC, Atwood JL, Barbour LJ, Croucher PD, Nichols PJ, Smith NO, Skelton BW, White AH, Raston CL (1999) J Chem Soc Dalton Trans: 2927
85. Croucher PD, Marshall JME, Nichols PJ, Raston CL (1999) Chem Commun: 193
86. Wang M-X, Zhang X-H, Zheng Q-Y (2004) Angew Chem Int Ed 43:838
87. Schuster DL, Rosenthal J, MacMahon S, Jarowski PD, Alabi CA, Guldi DM (2002) Chem Commun: 2538
88. Diederich F, Effing J, Jonas U, Jullien L, Plesnivy T, Ringsdorf H, Thilgen C, Weinstein D (1992) Angew Chem Int Ed Engl 31:1599
89. Tashiro K, Aida T, Zheng J-Y, Kinbara K, Saigo K, Sakamoto S, Yamaguchi K (1999) J Am Chem Soc 121:9477
90. Zheng J-Y, Tashiro K, Hirabayashi Y, Kinbara K, Saigo K, Aida T, Sakamoto S, Yamaguchi K (2001) Angew Chem Int Ed 40:1858
91. Ouchi A, Tashiro K, Yamaguchi K, Tsuchiya T, Akasaka T, Aida T (2006) Angew Chem Int Ed 45:3542
92. Shoji Y, Tashiro K, Aida T (2006) J Am Chem Soc 128:10690
93. Shoji Y, Tashiro K, Aida T (2004) J Am Chem Soc 126:6570
94. Tashiro K, Hirabayashi Y, Aida T, Saigo K, Fujiwara K, Komatsu K, Sakamoto S, Yamaguchi K (2002) J Am Chem Soc 124:12086
95. Tashiro K, Aida T (2001) J Incl Phenom Macrocycl Chem 41:215

96. Nishioka T, Tashiro K, Aida T, Zheng J-Y, Kinbara K, Saigo K, Sakamoto S, Yamaguchi K (2000) Macromolecules 33:9182
97. Yamaguchi T, Ishii N, Tashiro K, Aida T (2003) J Am Chem Soc 125:13934
98. Ayabe M, Ikeda A, Shinkai S, Sakamoto S, Yamaguchi K (2002) Chem Commun: 1032
99. Bhattacharya S, Tominaga K, Kimura T, Uno H, Komatsu N (2007) Chem Phys Lett 433:395
100. Sun D, Tham FS, Reed CA, Chaker L, Burgess M, Boyd PDW (2000) J Am Chem Soc 122:10704
101. Dudic M, Lhotak P, Stibor I, Petriclova H, Lang K (2004) New J Chem 28:85
102. Arimura T, Nishioka T, Suga Y, Murata S, Tachiya M (2002) Mol Cryst Liq Cryst 379:413
103. Wu Z-Q, Shao X-B, Li C, Hou J-L, Wang K, Jiang X-K, Li Z-T (2005) J Am Chem Soc 127:17460
104. Kubo Y, Sugasaki A, Ikeda M, Sugiyasu K, Sonoda K, Ikeda A, Takeuchi M, Shinkai S (2002) Org Lett 4:925
105. Ayabe M, Ikeda A, Kubo Y, Takeuchi M, Shinkai S (2002) Angew Chem Int Ed 41:2790
106. Bonifazi D, Spillmann H, Kiebele A, Wild M, Seiler P, Cheng F, Guntherodt J, Jung T, Diederich F (2004) Angew Chem Int Ed 43:4759
107. Guldi DM, Rahman GMA, Zerbetto F, Prato M (2005) Acc Chem Res 38:871
108. Liu J, Rinzler AG, Dai H, Hafner JH, Bradley RK, Boul PJ, Lu A, Iverson T, Shelimov K, Huffman CB, R-Macias F, Shon Y-S, Lee TR, Colbert DT, Smalley RE (1998) Science 280:1253
109. Curran SA, Ajayan PM, Blau WJ, Carroll DL, Coleman JN, Dalton AB, Davey AP, Drury A, McCarthy B, Maier S, Strevens A (1998) Adv Mater 10:1091
110. Chen J, Hamon MA, Hu H, Chen Y, Rao AM, Eklund PC, Haddon RC (1998) Science 282:95
111. Wang X, Liu Y, Qiu W, Zhu D (2002) J Mater Chem 12:1636
112. Basiuk EV, R-Akimova EV, Basiuk VA, A-Najarro D, Saniger JM (2002) Nano Lett 2:1249
113. Murakami H, Nomura T, Nakashima N (2003) Chem Phys Lett 378:481
114. Li H, Zhou B, Lin Y, Gu L, Wang W, Fernando KAS, Kumar S, Allard LF, Sun Y-P (2004) J Am Chem Soc 126:1014
115. Cheng F, Adronov A (2006) Chem Eur J 12:5053
116. Rahman GMA, Guldi DM, Campidelli S, Prato M (2006) J Mater Chem 16:62
117. Cheng F, Zhang S, Adronov A, Echegoyen L, Diederich F (2006) Chem Eur J 12:6062
118. Chen J, Collier CP (2005) J Phys Chem B 109:7605
119. Boul PJ, Cho D-G, Rahman GMA, Marquez M, Ou Z, Kadish KM, Guldi DM, Sessler JL (2007) J Am Chem Soc 129:5683
120. Hasobe T, Fukuzumi S, Kamat PV (2005) J Am Chem Soc 127:11884
121. Hasobe T, Fukuzumi S, Kamat PV (2006) J Phys Chem B 110:25477
122. Alvaro M, Atienzar P, Cruz P, Delgado JL, Troiani V, Garcia H, Langa F, Palkar A, Echegoyen L (2006) J Am Chem Soc 128:6626
123. Guldi DM (2005) J Phys Chem B 109:11432
124. Ehli C, Rahman GMA, Jux N, Balbinot D, Guldi DM, Paolucci F, Marcaccio M, Paolucci D, M-Franco M, Zerbetto F, Campidelli S, Prato M (2006) J Am Chem Soc 128:11222
125. Guldi DM, Rahman GMA, Jux N, Tagmatarchis N, Prato M (2004) Angew Chem Int Ed 43:5526
126. Nakashima N, Tomonari Y, Murakami H (2002) Chem Lett 31:638

127. Saito K, Troiani V, Qiu H, Solladie N, Sakata T, Mori H, Ohama M, Fukuzumi S (2007) J Phys Chem C 111:1194
128. Solladie N, Hamel A, Gross M (2000) Tetrahedron Lett 41:6075
129. Fujitsuka M, Hara M, Tojo S, Okada A, Troiani V, Solladie N, Majima T (2005) J Phys Chem B 109:33
130. Ortis-Acevedo A, Xie H, Zorbas V, Sampson WM, Dalton AB, Baughman RH, Draper RK, Musselman IH, Dieckmann GR (2005) J Am Chem Soc: 9512
131. Guldi DM, Taieb H, Rahman GMA, Tagmatarchis N, Prato M (2005) Adv Mater 17:871
132. Satake A, Miyajima Y, Kobuke Y (2005) Chem Mater 17:716
133. Tsang SC, Guo Z, Chen YK, Green MLH, Hill HAO, Hambley TW, Sadler PJ (1997) Angew Chem Int Ed Engl 36:2198
134. Guo Z, Sadler PJ, Tsang SC (1998) Adv Mater 10:701
135. Nakashima N, Okuzono S, Murakami H, Nakai T, Yoshikawa K (2003) Chem Lett 32:456
136. Zheng M, Jogota A, Semke ED, Diner BA, Mclean RS, Lustig SR, Richardson RE, Tassi NG (2003) Nat Mater 2:338
137. Zheng M, Jogota A, Strano MS, Santos AP, Barone P, Chou SG, Diner BA, Dresselhause MS, Mclean RS, Onoa GB, Samsonidze GG, Semke ED, Usrey M, Walls DJ (2003) Science 302:1545
138. Katz E, Willner I (2004) Chem Phys Chem 5:1084
139. Huang X, Mclean RS, Zheng M (2005) Anal Chem 77:6225
140. Strano MS, Zheng M, Jagota A, Onoa GB, Heller DA, Barone PW, Usrey ML (2004) Nano Lett 4:543
141. Dodziuk H, Ejchart A, Anczewski W, Ueda H, Krinichnaya E, Dolgonos G, Kutner W (2003) Chem Commun: 986
142. Fagan JA, Simpson JR, Bauer BJ, Lacerda SHDP, Becker ML, Chun J, Migler KB, Walker ARH, Hobbie EK (2007) J Am Chem Soc 129:10607
143. Heller DA, Jeng ES, Yeung T-K, Martinez BM, Moll AE, Gastala JB, Strano MS (2006) Science 311:508
144. Bae A-H, Hatano T, Nakashima N, Murakami H, Shinkai S (2004) Org Biomol Chem 2:1139
145. Zheng M, Diner BA (2004) J Am Chem Soc 126:15490
146. Zhao X, Johnson JK (2007) J Am Chem Soc 129:10438
147. Lu G, Maragakis P, Kaxiras E (2005) Nano Lett 5:897
148. Xin H, Woolley AT (2003) J Am Chem Soc 125:8710
149. Chen Y, Liu H, Ye T, Kim J, Mao C (2007) J Am Chem Soc 129:8696
150. Lin Y, Taylor S, Li H, Fernando KAS, Qu L, Wang W, Gu L, Zhou B, Sun Y-P (2004) J Mater Chem 14:527
151. Bianco A, Prato M (2003) Adv Mater 15:1765
152. Bianco A, Kostarelos K, Partidos CD, Prato M (2005) Chem Commun: 571
153. Pantarotto D, Briand J-P, Prato M, Bianco A (2004) Chem Commun: 16
154. Pantarotto D, Singh R, McCarthy D, Erhardt M, Briand J-P, Prato M, Kostarelos K, Bianco A (2004) Angew Chem Int Ed 43:5242
155. Singh R, Pantarotto D, McCarthy D, Chaloin O, Hoebeke J, Partidos CD, Briand J-P, Prato M, Bianco A, Kostarelos K (2005) J Am Chem Soc 127:4388
156. Bianco A, Hoebeke J, Godefroy S, Chaloin O, Pantarotto D, Briand J-P, Muller S, Prato M, Partidos CD (2005) J Am Chem Soc 127:58
157. Kam NWS, O'Connell M, Wisdom JA, Dai H (2005) PNAS 102:11600
158. Kam NWS, Liu Z, Dai H (2005) J Am Chem Soc 127:12492
159. Jeng ES, Moll AE, Roy AC, Gastala JB, Strano MS (2006) Nano Lett 6:371

160. Williams KA, Veenhuizen PTM, Torre BG, Eritja R, Dekker C (2002) Nature 420:761
161. Li J, Ng HT, Cassell A, Fan W, Chen H, Ye Q, Koehne J, Han J, Meyyappan M (2003) Nano Lett 3:597
162. Ito T, Sun L, Crooks RM (2003) Chem Commun: 1482
163. Gao H, Kong Y, Cui D (2003) Nano Lett 3:471
164. Tsang SC, Davis JJ, Hill HAO, Leung YC, Sadler PJ (1995) Chem Commun: 1803
165. Tsang SC, Davis JJ, Hill HAO, Leung YC, Sadler PJ (1995) Chem Commun: 2579
166. Davis JJ, Green MLH, Hill HAO, Leung YC, Sadler PJ, Sloan J, Xavier A, Tsang SC (1998) Inorg Chim Acta 272:261
167. Davis JJ, Coles RJ, Hill HAO (1997) J Elecreoanal Chem 440:279
168. Balavoine F, Schultz P, Richard C, Mallouh V, Ebbesen TW, Mioskowski C (1999) Angew Chem Int Ed 38:1912
169. Erlanger BF, Chen B-X, Zhu M, Brus L (2001) Nano Lett 1:465
170. Azamian BR, Davis JJ, Coleman KS, Bagshaw CB, Green MLH (2002) J Am Chem Soc 124:12664
171. Shim M, Kam NWS, Chen RJ, Li Y, Dai H (2002) Nano Lett 2:285
172. Chen RJ, Zhang Y, Wang D, Dai H (2001) J Am Chem Soc 123:3838
173. Kam NWS, Dai H (2005) J Am Chem Soc 127:6021
174. Kam NWS, Jessop TC, Wender PA, Dai H (2004) J Am Chem Soc 126:6850
175. Chen RJ, Bangsaruntip S, Drouvalakis KA, Kam NWS, Shim M, Li Y, Kim W, Utz PJ, Dai H (2003) PNAS 100:4984
176. Satishkumar BC, Brown LO, Gao Y, Wang C-H, Wang H-L, Doorn SK (2007) Nat Nanotechnol 2:560
177. Besteman K, Lee J, Wiertz FGM, Heering HA, Dekker C (2003) Nano Lett 3:727
178. Hirata T, Amiya S, Akiya M, Takei O, Sakai T, Hatakeyama R (2007) Appl Phys Lett 90:233106
179. Wang S, Humphreys ES, Chung S-Y, Delduco DF, Lustig SR, Wang H, Parker KN, Rizzo NW, Subramoney S, Chiang Y-M, Jagota A (2003) Nat Mater 2:196
180. Dieckmann GR, Dalton AB, Johnson PA, Razal J, Chen J, Giordano GM, Munoz E, Musselman IH, Baughman RH, Draper RK (2003) J Am Chem Soc 125:1770
181. Zorbas V, Ortis-Acevedo A, Dalton AB, Yoshida MM, Dieckmann GR, Draper RK, Jose-Yacaman M, Baughman RH, Musselman IH (2004) J Am Chem Soc 126:7222
182. Ortis-Acevedo A, Dieckmann GR (2004) Tetrahedron Lett 45:6795
183. Braun T (1997) Fullerene Sci Technol 5:615
184. Andersson T, Nilsson K, Sundahl M, Westman G, Wennerstrom O (1992) Chem Commun: 604
185. Andersson T, Westman G, Wennerstrom O, Sundahl M (1994) Perkin Trans 2:1097
186. Andersson T, Westman G, Stenhagen G, Sundahl M, Sundahl M (1995) Tetrahedron Lett 36:597
187. Andersson T, Sundahl M, Westman G, Wennerstrom O (1994) Tetrahedron Lett 35:7103
188. Yoshida Z, Takekuma H, Takekuma S, Matsubara Y (1994) Angew Chem Int Ed Engl 33:1597
189. Murthy CN, Geckeler KE (2001) Chem Commun: 1194
190. Priyadarsini KI, Mohan H, Tyagi AK, Mittal JP (1994) J Phys Chem 98:4756
191. Wang G-W, Komatsu K, Murata Y, Shiro M (1997) Nature 387:583
192. Braun T, B-Barcza A, Barcza L, K-Thege I, Fodor M, Migali B (1994) Solid State Ionics 74:47
193. Komatsu K, Fujiwara K, Murata Y, Braun T (1999) Perkin Trans 1:2963
194. Priyadarsini KI, Mohan H, Mittal JP (1995) Fullerene Sci Technol 3:479

195. Sundahl M, Andersson T, Nilsson K, Wennerstom O, Westman G (1993) Synth Metal 55–57:3252
196. Masuhara A, Fujitsuka M, Ito O (2000) Bull Chem Soc Jpn 73:2199
197. Boulas P, Kutner W, Jones MT, Kadish KM (1994) J Phys Chem 98:1282
198. Takekuma S, Takekuma H, Matsumoto T, Yoshida Z (2000) Tetrahedron Lett 41:4909
199. Ikeda A, Sato T, Kitamura K, Nishiguchi K, Sasaki Y, Kikuchi J, Ogawa T, Yogo K, Takeya T (2005) Org Biomol Chem 3:2907
200. Chen J, Dyer MJ, Yu M-F (2001) J Am Chem Soc 123:6201
201. Chambers G, Carroll C, Farrell GF, Dalton AB, McNamara M, Panhuis M, Byrne HJ (2003) Nano Lett 3:843
202. Komatsu K, Fujiwara K, Murata Y, Braun T (1999) Perkin Trans 1:2963
203. Ikeda A, Hayashi K, Konishi T, Kikuchi J (2004) Chem Commun: 1334
204. Star A, Steuerman DW, Heath JR, Stoddart JF (2002) Angew Chem Int Ed 41:2508
205. Kim O-K, Je J, Baldwin JW, Kooi S, Pehrsson PE, Buckley LJ (2003) J Am Chem Soc 125:4426
206. Bandyopadhyyaya R, N-Roth E, Regev O, Y-Rozen R (2002) Nano Lett 2:25
207. Yang H, Wang SC, Mercier P, Akins DL (2006) Chem Commun: 1425
208. Chattopadhyay D, Galeska I, Papadimitrakopoulos F (2003) J Am Chem Soc 125:3370
209. Maeda Y, Kimura S, Kanda M, Hirashima Y, Hasegawa T, Wakahara T, Lian Y, Nakahodo T, Tsuchiya T, Akasaka T, Lu J, Zhang X, Gao Z, Yu Y, Nagase S, Kazaoui S, Minami N, Shimizu T, Tokumoto H, Saito R (2005) J Am Chem Soc 127:10287
210. Maeda Y, Kanda M, Hirashima Y, Hasegawa T, Kimura S, Lian Y, Wakahara T, Akasaka T, Kazaoui S, Minami N, Okazaki T, Hayamizu Y, Hata K, Lu J, Nagase S (2006) J Am Chem Soc 128:12239
211. Moulton SE, Maugey M, Poulin P, Wallace GG (2007) J Am Chem Soc 129:9452
212. Health JR, O'Brien SC, Zhang Q, Liu Y, Curl RF, Kroto HW, Tittel FK, Smalley RE (1985) J Am Chem Soc 107:7779
213. Nagase S, Kobayashi K, Akasaka T (1996) Bull Chem Soc Jpn 69:2131
214. Bethune DS, Johnson RD, Salem JR, Vries MS, Yannoi CS (1993) Nature 366:123
215. Liu S, Sun S (2000) J Organomet Chem 599:74
216. Shinohara H (2000) Rep Prog Phys 63:843
217. Akasaka T, Nagase S (2002) Endofullerenes. Kluwer Academic Publishers, Dordrecht
218. Chikkannanavar SB, Smith BW, Luzzi DE (2006) In: O'Connell M (ed) Carbon Nanotubes. Taylor & Francis, Boca Raton, FL, p 51
219. Ajayan PM, Iijima S (1993) Nature 361:333
220. Seraphin S, Zhou B, Jiao J, Withers JC, Louftfy R (1993) Nature 362:503
221. Tsang SC, Chen YK, Harris PJF, Green MLH (1994) Nature 372:159
222. Vostrowsky O, Hirsch A (2004) Angew Chem Int Ed 43:2326
223. Sloan J, Kirkland AI, Hutchinson JL, Green MLH (2002) Acc Chem Res 35:1054
224. Smith BW, Monthioux M, Luzzi DE (1998) Nature 396:323
225. Khlobystov AN, Britz DA, Briggs GAD (2005) Acc Chem Res 38:901
226. Hornbaker DJ, Kahng S-J, Misra S, Smith BW, Johnson AT, Mele EJ, Luzzi DE, Yazdani A (2002) Science 295:828
227. Suenaga K, Tence M, Mory C, Colliex C, Kato H, Okazaki T, Shinohara H, Hirahara K, Bandow S, Iijima S (2000) Science 290:2280
228. Khlobystov AN, Porfyrakis K, Kanai M, Britz DA, Ardavan A, Shinohara H, Dennis TJS, Briggs GAD (2004) Angew Chem Int Ed 43:1386
229. Suenaga K, Taniguchi R, Shimada T, Okazaki T, Shinohara H, Iijima S (2003) Nano Lett 3:1395
230. Stercel F, Nemes NM, Fischer JE, Luzzi DE (2002) Mater Res Soc Symp Proc 706:245

231. Li L-J, Khlobystov AN, Wiltshire JG, Briggs GAD, Nicholas RJ (2005) Nat Mater 4:481
232. Morgan DA, Sloan J, Green MLH (2002) Chem Commun: 2442
233. Koshino M, Tanaka T, Solin N, Suenaga K, Isobe H, Nakamura E (2007) Science 316:853
234. Yanagi K, Miyata Y, Kataura H (2006) Adv Mater 18:437
235. Yanagi K, Iakoubovskii K, Matsui H, Matsuzaki H, Okamoto H, Miyata Y, Maniwa Y, Kazoui S, Minami N, Kataura H (2007) J Am Chem Soc 129:4992
236. Kondratyuk P, Yates JT (2007) J Am Chem Soc 129:8736
237. Nishide D, Wakabayashi T, Sugai T, Kitaura R, Kataura H, Achiba Y, Shinohara H (2007) J Phys Chem C 111:5178
238. Takenobu T, Takano T, Shiraishi M, Murakami Y, Ata M, Kataura H, Achiba Y, Iwasa Y (2003) Nat Mater 2:683
239. Chen S, Wu G, Sha M, Huang S (2007) J Am Chem Soc 129:2416
240. Kataura H, Maniwa Y, Abe M, Fujiwara A, Kodama T, Kikuchi K, Imahori H, Misaki Y, Suzuki S, Achiba Y (2002) Appl Phys A 74:349
241. Schulte K, Swarbrick JC, Smith NA, Bondino F, Magnano E, Khlobystov AN (2007) Adv Mater 19:3312
242. Nishide D, Dohi H, Wakabayashi T, Nishibori E, Aoyagi S, Ishida M, Kikuchi S, Kitaura R, Sugai T, Sakata M, Shinohara H (2006) Chem Phys Lett 428:356
243. Suzuki T, Nakashima K, Shinkai S (1994) Chem Lett: 699
244. Atwood JL, Koutsantonis GA, Raston CL (1994) Nature 368:229

Subject Index

Ag$^+$ cation recognition 8
Anion recognition, designed heterocycles 14, 25
–, –, lanthanide complexes 27
Anion sensing 29, 32
Anthracene rings, crownophanes 56
Azacalix[2]arene[2]pyridine 167
Azacalix[2]arene[2]triazine 83
Azacalix[3]arene 82
Azacalix[4]arene 82
Azacalix[4]arene[4]pyridine 85, 167
Azacalix[5]arene 85
Azacalix[6]arene 85
Azacalix[8]arene 85
Azacalix[10]arene 87
Azacalixarenes 73
–, inclusion properties 90
–, structural investigations 82
–, syntheses 75
Azacalixpyridines 76
Azacrown ether derivatives 33

Benzene rings, crownophanes 46
Bicyclo[2.2.2]octadiene 151
Big receptor–big guest complexes 12
Big receptor–small guest complexes 12
Bipyridine rings, crownophanes 59
Bipyridinocrownophanes 59
Bovine serum albumin 185
1-Butyl-3-methylimidazolium hexafluorophosphate [bmim][PF6] 189

Calixarenes 7, 73
Calixfurans 97, 113
–, reactions 108
Calix[5]furans, syntheses 105
Calix[6]furans, syntheses 105
Calix[4]furans, syntheses 99
Calix[4]imidazolium[2]pyridine 19
Calixpyrroles 22, 98
Carbohydrates, supramolecules, carbon nanotubes 187
–, –, fullerenes 186
Carbon nanotubes 161
–, endohedral supramolecular chemistry 189
–, exohedral supramolecular chemistry 167
o-Carboranes 189
Catenane 2, 43
Cation recognition, designed heterocycles 6
–, supramolecular 11
Chirality 1
Chirality optimized receptors 9
Chitosan 188
Chlorophylls 125
CNT/DNA hybrids 183
Complexation 161
Convergent fragment coupling synthesis 77
Crown ethers 12, 14, 43
Crownoanthracenophane 56
Crownopaddlanes 50
Crownophanes 43
–, heteroaromatic ring 58
Crownopyrenophanes 57
Crownopyridinophanes 58
Cryptands 14
Cyclen–lanthanide complexes 32
Cycloaddition reactions 112
Cyclodextrins 187
Cyclophane 17, 43, 62
Cyclotriveratrylene 162
Cytochrome c 13, 185

Dipyrrolylquinoxaline 25
DNA, supramolecules with carbon
　nanotubes 183
DWNTs 163

Electronic devices 123

Fluorenone rings, crownophanes 55
Foldamers 5
Fullerenes 5, 139, 162
–, encapsulation 169
–, endohedral supramolecular chemistry
　189
–, exohedral supramolecular chemistry
　167
–, non-covalent functionalization 161
–, peapods 189
Furfuryl alcohol 100
Furylpyrrolylmethane 104

Guanosine derivatives 5
Gum arabic 188

Hemoglobin proteins 3
Heterocalixarenes 7, 97
–, transformations 108
Heterocycle–lanthanide complexes, anion
　recognition/sensing 28
Heterocycles, cation recognition 6
Heterocyclic receptors, cationic 17
–, neutral 20
Hyaluronic acid 189
Hydrogen-bonding interactions 127

Imidazoles 15
Imidazolium receptors, cyclic 19
Incident photon-to-current conversion
　efficiency (IPCE) 137
Inclusion properties, calixfurans 113, 117

Lanthanide complexes 1
–, luminescence sensing 29
–, NMR sensing 35
Lasalocid ionophore 9
Luminescence 1
Luminescence sensing, lanthanide
　complexes 29

Metacyclophane skeleton 73
Metal tetraazaannulenes 167

Metal–ligand coordinative interactions
　126
Metallocycle 2
Metallofullerenes 189
Molecular recognition 1
–, heterocycles, lanthanide complexes 1
MWNTs 163

Naphthalene rings, crownophanes 53

Oligopyridine derivatives 2
Optical devices 123
Organic field-effect transistors (OFETs)
　148
Organic light-emitting diodes (OLED)
　149
π–π interactions 123, 139
–, stacking 123

Peptides, supramolecules with carbon
　nanotubes 185
Phenanethroline rings, crownophanes
　60
Phenanthridinium 34
Phospholipid-PEG-X 183
Photochemical energy conversion 137
Photoconversion systems 143
Photosynthesis, artificial 137
Phthalocyanines 123
–, supramolecules 124
Polyaromatic rings, crownophanes 55
Polyketide synthases 3
Porphyrin/methacrylic acid 182
Porphyrin/pyrene 182
Porphyrins 123, 161
–, monomers 167, 172
–, –, covalently linked oligomers for TPA
　131
–, nanoring 127
–, oligomers 169, 178
–, –, TPA 133
–, supramolecules 124
–, –, analogues with carbon nanotubes
　172
–, –, analogues with fullerenes 167
Protein A 185
Proteins, supramolecules with carbon
　nanotubes 185
Pyrene rings, crownophanes 57
Pyrenophanes 57

Subject Index

Pyridine rings, crownophanes 58
Pyridinium tetramer 17
Pyridinocrownophanes 59
Pyridothioxanthone 34
Pyrrolylamideurea 25

Quinoline chromophore 14

Receptors 1
–, geometry optimized 7
Rotaxane 43, 62
Salmonella typhimurium, Cl⁻ anion channel 3

Sapphyrin 21, 174
Small receptor–big guest complexes 11
Small receptor–small guest complexes 11
Solar cells, dye-sensitized 137
Solution-processed tetrabenzoporphyrins (TBPs), OFETs 151
Stilbene rings, crownophanes 55
Streptavidine 185
SWNTs 163

TBP(Cu) thin film 153
Tetracyanoquinodimethane 189
Tetrafuluorotetracyanoquinodimethane 189
Tetrakis(4-sulfonatophenyl)porphyrin 136, 174
Tetramethyltetraselenafulvalene 189
Tetraoxapaddlane 56
Tetraoxaporphyrinogens 98
Tetraphenylporphyrins (TPPs) 139
Tetrathiafulvalene 189
Thiacalixarenes 73
Thiacrownophanes 50
Thiamine (vitamin B_1) 17
Thin-film organic photovoltaic cells 137
Thiophene 97
Tripyridinium receptor 20
Tris(2-pyridylmethyl)amines 9
Two-photon absorption (TPA) 131

Urea 16

van der Waals interaction 123
Viral coat assemblies 3

Zinc chlorophyll 125
Zinc diphenylporphyrin 190
Zinc imidazolylporphyrin 143
Zinc porphyrins 5

Printing: Krips bv, Meppel, The Netherlands
Binding: Stürtz, Würzburg, Germany